Electronics for Higher TEC

Electronics for Higher TEC

S. A. Knight

GRANADA
London Toronto Sydney New York

Granada Publishing Limited–Technical Books Division
Frogmore, St Albans, Herts AL2 2NF
and
36 Golden Square, London W1R 4AH
515 Madison Avenue, New York, NY 10022, USA
117 York Street, Sydney, NSW 2000, Australia
60 International Boulevard, Rexdale, Ontario R9W 6J2, Canada
61 Beach Road, Auckland, New Zealand

British Library Cataloguing in Publication Data
Knight, S.A.
Electronics for higher TEC.
1. Electronics
I. Title
621.381   TK7815
ISBN 0-246-11559-9

First published in Great Britain by Granada Publishing Ltd 1983

Printed in Great Britain by Richard Clay (The Chaucer Press) Ltd,
Bungay, Suffolk

Granada ®
Granada Publishing ®

Contents

Preface

This book covers the relevant TEC units in Electronics at level IV
together with part of the requirements for level V. The contents are
a direct development from level III Electrical and Electronic Principles and it is assumed that this level, or its equivalent, has been
adequately covered. A note and a qualification about the structure
of the book will not be amiss.

The TEC guide syllabus for levels IV and V in Electronics is very
wide and although it covers much of the contents of the City & Guilds
Higher Technician and the H.N.C. A.1 syllabuses, which it replaces,
its presentation is different and it is not possible clearly to define a 'set-piece' coverage in the way that these earlier programmes
did. Because of this it did not prove an easy task to write a book
which would cover adequately a unit at level IV and level V in the
limited space available. The approach then has been to cover the
hard core, as it were, of electronics materials and systems as are
needed as a minimum requirement by technician students, and to dress
this skeleton with selected topics which are of general relevance
to the syllabus overall. In this way, a common groundwork of the
essential electronics has (hopefully) been provided in a single and
not encyclopedic volume.

The associated mathematics, unavoidable as they are, are not
severe; only sufficient mathematics to support the theory as it
develops are included, and circuit analysis is treated as far as
possible in a qualitative way in addition to the mathematics. A particular feature which supports this approach is the emphasis placed
on illustrative examples fully worked out and explained throughout

the text, a number of qualitative and quantitative problems for self-assessment given at the end of each chapter, and references to basic experimental work at a number of points throughout the text.

S. A. Knight

1 Conductors, semiconductors and junctions

1.1 CONDUCTORS AND INSULATORS

Conduction occurs in vacuums, gases, liquids and solids if carriers are available to transfer charge under the influence of an applied field. In a vacuum, conduction is possible if free electrons are available as in thermionic valves and cathode-ray tubes. In an ionized gas, charged ions as well as electrons provide the necessary carriers. Positive and negative ions are the charge carriers in a liquid. Solids vary widely in the type and number of available carriers. Insulating materials have practically no available charge carriers; conductors have large numbers of such carriers; and semiconductors have numbers of carriers which are intermediate between those of insulators and conductors.

1.2 METALLIC CONDUCTORS

In a crystalline solid, the atoms forming each crystal are held together in a systematic array or lattice by chemical bonds. These bonds consist of electron pairs which are shared by adjacent atoms. Electrons in bonds are in a lower energy state than would be the case if the bond was broken, since work has to be done to break apart the bond and hence to separate the atoms so bonded. When water freezes, for example, it gives off heat. To melt the ice this same heat must be added. The energy represented by this much heat is just the energy needed to break all the chemical bonds between the atoms in an element or between the molecules in a compound.

In a metal such as copper, the bonds are formed by certain, but not all, of the outermost electrons; others are bound to the nucleus

so lightly that they do not participate in the bonds but are attracted
to numerous neighbouring nuclei, not being associated with any particular nucleus. In fact, there are about as many free electrons in a
given volume as there are atoms. These conduction electrons are
available, therefore, to drift randomly throughout the solid as isolated charges. It is these negatively charged particles which form
the carriers of electric current by drifting in a particular direction when an electric field is applied, see Fig. 1.1. At absolute
zero temperature $0^{\circ}K$ (= $-273.16^{\circ}C$), these carriers encounter no
opposition to motion and the resistance of the metal is zero.

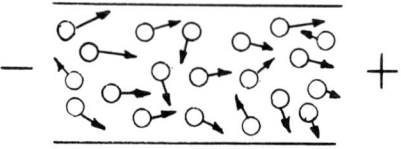

Figure 1.1

At ordinary (room) temperatures, those atoms (ions) that are
deficient in their outermost electrons (or electron) possess kinetic
energy in the form of vibration about a neutral position in the
crystalline structure. This vibrational energy results from the
effect of heat and is therefore temperature dependent. Free carriers
moving throughout the lattice continually encounter the fields of
the vibrating ions and there is a continual interchange of energy.
Unless a field is applied the resulting motion of the carriers is
random, the net current is zero and the net voltage established
across any part of the conductor is also zero. As temperature increases, the random motion increases, the time between collisions
decreases and the conductivity of the metal falls. It is a characteristic of metals that resistance (the reciprocal of conductivity)
increases with temperature.

1.3   SEMICONDUCTORS

In semiconducting materials ALL the outer electrons of each atom are
involved in chemical bonding and free electrons in a perfect crystal
lattice would be non-existent. The two semiconductors of importance
in solid state electronics, silicon and germanium, belong to the

fourth column of the periodic table, although other semiconductors involving compounds of elements from the second and fourth columns, such as cadmium sulphide, have important applications. The crystal structure of silicon or germanium is therefore tetrahedral, with each atom sharing one outermost (valence) electron with each of four neighbouring atoms. Nothing essential is lost and clarity is gained when the structure is considered to lie all in one plane as shown in Fig. 1.2. Here the circles represent the nuclei and the inner shells of the atoms with the covalent bonds shown as parentheses. In pure silicon or germanium at the absolute zero of temperature each bond would be complete, there would be no free electrons, and the crystal would be an insulator.

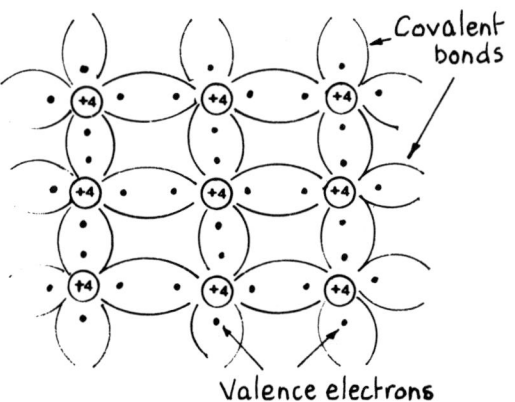

Figure 1.2

The energy that an electron has is the sum of its potential and kinetic energies. An electron, by virtue of its position, can have energy even when at rest. Electrons in the valence bonds have low potential energy; additional energy is required to free an electron from a bond. By the same token, a free electron is in a relatively high energy state. In addition to being free, an electron can have theoretically unlimited kinetic energy, but to be free it must have at LEAST a certain total energy.

In a pure crystal there is a range of total energies which the valence electrons may have. This range is the 'valence' energy band. There is an open range of energies above a certain minimum level which free electrons may have. This range is the 'conduction' band because, as we have seen, such electrons are charge carriers. Between the top of the valence band i.e. the MOST energy a bound electron can have, and the bottom of the conduction band i.e. the LEAST energy a free electron can have, there may be a gap representing a range of energies that no electron can have. Conductors and insulators can be explained in terms of such energy bands; a good conductor is characterised by an energy band system in which a large number of electrons have energies which reside in the conduction band; a good insulator is one in which there are no electrons with energies which place them in the conduction band and moreover a very wide gap exists between the highest valence band and the conduction band.

The energy band pictures for conductors, insulators and semiconductors are shown in Fig. 1.3. In semiconductor materials an electron, in order to escape a valence bond and become a free carrier, must acquire at least enough energy to cross the gap. This cannot be done by instalments so to speak. An electron cannot acquire part of the required energy at some particular time and hang about for the rest to arrive later. It must either acquire the minimum energy necessary to cross the gap in one package or stay where it is.

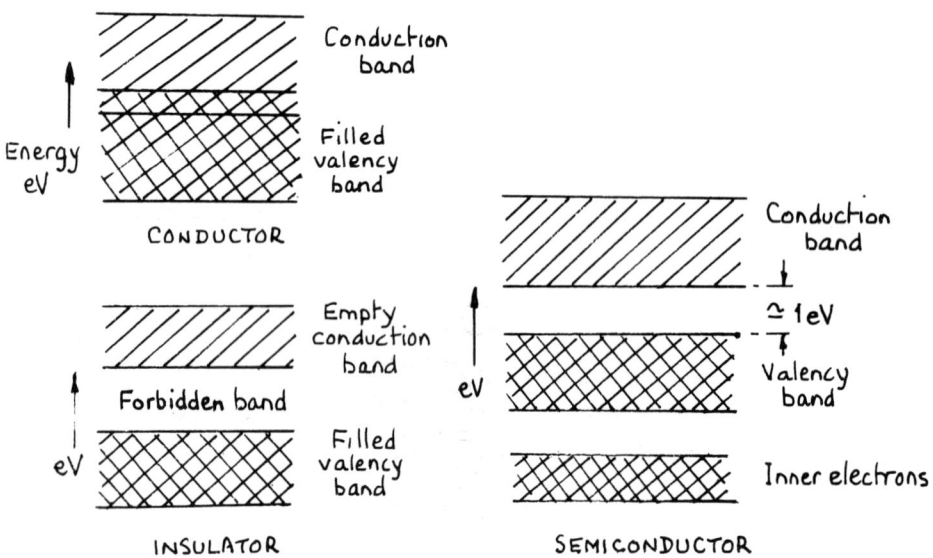

Figure 1.3

4

The energy required to make the crossing (or to break a covalent bond) is about 1.1 eV for silicon and about 0.7 eV for germanium. An electron-volt (eV) is the energy an electron acquires in accelerating through a potential difference of 1 V, about 1.6×10^{-19} joule. At ordinary temperatures some electrons acquire enough thermal energy to cross the gap and become free electrons. However, the distribution of thermal energies at room temperature is such that for each 50 meV increase in the effective gap 'width' the number of thermally excited free electrons decreases by about one-third.

When an electron acquires enough energy to break out of a valence bond it becomes free to drift under the action of an electric field and so conduct electricity. The bond vacated by the electron is now a half filled bond with only one electron present rather than two, and it too is now free to wander around the crystal and behave as a charge carrier. The region in which such a vacancy exists has a net positive charge which is equal to the net negative charge $e(= -1.6 \times 10^{-19}$C) carried by the departed electron. Figure 1.4 shows a silicon crystal with a broken covalent bond. In such semi-conductors both electrons and vacancies (holes) contribute to electrical conduction, electrons moving in the conduction band and holes in the valence band. If a valence electron from an adjacent covalent bond fills the hole (without having sufficient energy to escape into the conduction band), the hole appears in a new place and the effect is as if a charge of magnitude +e has moved to another point in the lattice. For reasons that we shall come to later, holes are not so mobile as electrons, that is, they drift at a slower rate under an applied field and they move in the opposite direction to the

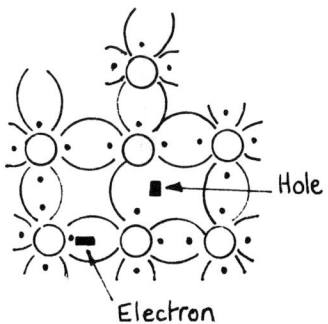

Figure 1.4

5

electrons. They also wander around under the influence of thermal agitation. Hence each ionized bond produces two carriers; one electron and one hole. Since electrons and holes are created at the same instants of time, a pure semiconductor is characterised by having exactly equal numbers of holes and free electrons, or hole-electron pairs as they are known. Hole-electron pairs are being generated continuously in a piece of semiconductor material and under conditions of thermal equilibrium they disappear by the process of recombination at the same rate, that is, electrons fill holes and both forms of carrier effectively vanish. At room temperature, only a very small fraction of the valence band electrons break into the conduction band and the semiconductor behaves as a very poor conductor. Conduction which takes place as a result of thermally generated hole-electron pairs is known as 'intrinsic' conduction and such material is known as intrinsic semiconductor. The energy gaps, intrinsic carrier density, conductivities and mobilities for silicon and germanium are given in Table 1.1.

Table 1.1

	Silicon	Germanium
Energy gap (eV)	1.1	0.7
Intrinsic carrier density ($/m^3$)	1.5×10^{16}	2.4×10^{19}
Conductivity (S/m)	4.5×10^{-4}	2.3
Hole mobility (m^2/V-s)	0.05	0.19
Electron mobility (m^2/V-s)	0.135	0.4

1.3.1 Doping

As a crystal of silicon or germanium is growing, the atoms at the cooling crystal surface lose energy, the electrons form their bonds with the atoms already attached and the new atoms take their place in the lattice. If impurities are present in the melt (as they invariably are), an occasional impurity atom of some other element will replace one of the silicon or germanium atoms and the resulting crystal will contain a number of such impurity atoms in its structure.

If the impurity is an element from the adjacent fifth column of the periodic table such as phosphorus, arsenic or antimony, it will

have five valence electrons and the effective ionic charge is +5e.
Thus after the four covalent bonds have been formed with the adjacent
silicon or germanium atoms, the remaining electron, with the addition
of a small amount of thermal energy, becomes a free electron available for conduction. Thus the impurity centres are almost completely
ionized at room temperature and each column 5 atom 'donates' a free
electron to the crystal. Therefore all the extra electrons associated with the impurity are free from their parent atoms and a concentration of electrons equal to the impurity concentration is released into
the lattice; the conductivity of the crystal will consequently increase. Pentavalent impurity atoms of this kind are called 'donors'
and the crystal formed with them in occupation is called n-type semiconductor because of the presence of negative charge carriers in an
electrically neutral structure. The net positive charges left behind
are bound charges and being immobile cannot contribute to conduction.

Column 3 elements such as boron, aluminium, gallium and indium
have three valence electrons, so when they replace silicon (or
germanium) atoms in the lattice only three of the four required
covalent bonds can be completed. The unfilled bond constitutes a
hole. Corresponding to this missing electron there is one less positive charge on the impurity nucleus. When an electron from elsewhere
fills the hole and the hole effectively wanders off, the region
around the impurity is left with a net negative charge. This net
negative charge is immobile but the hole is mobile and current conduction can take place by way of positive charges. Because such trivalent impurities can accept an electron they are called 'acceptor'
atoms and because the material they occupy has an excess of holes
(or positive carriers) it is called p-type. In n-type material the
electrons are the 'majority' carriers, the holes are 'minority'
carriers. In p-type material the holes are the majority carriers
whilst the electrons are the minority carriers. The conductivity of
a semiconductor can be greatly increased, and to a precisely controlled extent, by adding small quantities of impurity elements to
the intrinsic material. This process is called doping.

1.3.2 Mass action and charge conservation
If doping with a pentavalent impurity increases the density of

electrons and doping with trivalent impurity increases the density
of holes, what happens to the densities of the minority carriers? We
can look into this by considering that since the appearance of every
free electron creates a hole, the birth rate for electrons and holes
must be the same. Also, since the only way they disappear is by recombination with each other, the death rates must be equal. If the
population is to remain stable in conditions of equilibrium, the
birth and death rates must be the same. Hence the rate of generation
equals the rate of recombination. Since the rate of generation is
independent of the population while the rate of recombination
'increases' with population, the population adjusts itself until the
rates are equal. Let:

R = rate of generation and recombination in electron-hole
pairs per second per cubic metre.

p = equilibrium density of holes per m^3

n = equilibrium density of electrons per m^3

T_p = average lifetime of the holes (seconds)

T_n = average lifetime of the electrons (seconds)

n_i = the intrinsic concentration in electron-hole pairs per
cubic metre.

Each density must be the product of the birth rate and lifetime,
hence

$$p = T_p R$$
$$n = T_n R \tag{1.1}$$

Notice that the ratio of lifetimes is the same as the ratio of concentrations. Now the lifetime of either carrier is inversely proportional to the concentration of the other, so we can write

$$T_p = \frac{K}{n} \quad \text{or} \quad T_n = \frac{K}{p}$$

where K is a proportionality constant which may or may not be a
function of temperature. Substituting into (1.1) we have from
either equation

$$pn = KR \tag{1.2}$$

Now even in a crystal which has been heavily doped, the great bulk of the atoms are still silicon (or germanium) and the thermal generation rate is unchanged from the intrinsic value, R. Also, in pure material $p = n = n_i$ hence

$$n_i^2 = KR$$

and

$$pn = n_i^2 \qquad (1.3)$$

This is the 'mass action' law which states that increasing the concentration of either carrier depresses that of the other, the product remaining the same as for the pure material.

That the product pn is constant is consistent with the idea of generation and recombination. If an electron density, for example, is increased by the introduction of donors, the recombination rate is enhanced but the generation rate is unchanged, hence the net density of the minority carriers falls.

When a donor has given up its electron, there is a bound positive charge left in the nucleus. Likewise when an acceptor has taken an electron into its bond the nucleus is short by one positive charge compared with the bound electrons around it, so in effect there is a bound negative charge concentrated at the nucleus. There are two types of charge to consider, therefore: the immobile ions and the mobile carriers. Let the immobile positive and negative ions be present in concentrations N_d and N_a, the subscripts indicating donors and acceptors. Since the total crystal must be electrically neutral, the sample will exhibit no net charge if

$$p + N_d = n + N_a$$

or

$$p - n = N_a - N_d \qquad (1.4)$$

By substitution and transformation between equations (1.3) and (1.4)

9

we find

$$p = \sqrt{n_i^2 + \left[\frac{N_a - N_d}{2}\right]^2} + \left[\frac{N_a - N_d}{2}\right]$$
(1.5)

$$n = \sqrt{n_i^2 + \left[\frac{N_a - N_d}{2}\right]^2} - \left[\frac{N_a - N_d}{2}\right]$$
(1.6)

From these by addition

$$n + p = \sqrt{4n_i^2 + [N_a - N_d]^2}$$
(1.7)

These equations have assumed that $N_a - N_d > 1$. If $N_a - N_d < 1$, p and n should be interchanged as should the subscripts a and d.

It is important to notice in these equations that it is only the EXCESS of one doping agent over the other that is significant. This is a fact of great practical importance for if it was otherwise the manufacture of junctions and transistors would be almost impossible.

In n-type material, created by the introduction of donor atoms into the intrinsic material, $N_a = 0$, hence from (1.4) the majority carrier density (n_n) is the sum of the donor density N_d and the intrinsic density n_i, that is

$$n_n = p + N_d$$

But $N_d \gg p$ in practical cases, hence

$$n_n \simeq N_d$$

Also, from (1.3) the density of holes in n-type material is

$$P_n = \frac{n_i^2}{n_n} \simeq \frac{n_i^2}{N_d}$$

In p-type material created by the addition of acceptor impurity to the intrinsic material, corresponding equations can be deduced and are

$$P_p \simeq N_a$$

$$n_p \simeq \frac{n_i^2}{N_a}$$

Thus under appreciable doping the majority carrier density approaches the excess density of doping centres, while the minority carrier density is depressed by the law of mass action. The effect of doping concentrations can be predicted using these equations.

1.4 CONDUCTIVITY AND MOBILITY

If a uniform electric field of intensity ε (V/m) is applied across the ends of a metal conductor, the electron charge carriers have superimposed on their normal random motion a drift component of velocity directed towards the positive pole of the field. Any particular electron loses its kinetic energy when it collides with an ion, but it then accelerates again under the action of the field and in turn again loses its energy at the next collision, see Fig. 1.5.

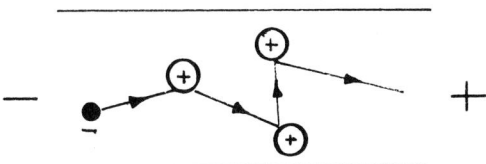

Figure 1.5

The time between collisions is a function of the random velocity and the length of the mean free path. The drift velocity directed towards the positive pole is on average directly proportional to ε since the acceleration experienced is constant for an average time interval. Hence

$$\bar{u} = \mu\varepsilon \quad \text{or} \quad \mu = \frac{\bar{u}}{\varepsilon}$$

where $\bar{u}$ is the mean drift velocity and the constant of proportionality μ is the mobility. Mobility is defined as the carrier velocity (m/s) imparted by unit field intensity (1 V/m). Its units are therefore metres per second/volts per metre or m^2/V-m. Carrier mobility is a property of the material; in copper it is about 3×10^{-2} $m^2/V\text{-}s$.

If there are n free electrons per cubic metre of material, each electron carrying charge e, the resulting current has a density J (A/m^2) given by

$$J = ne\mu = ne\mu\varepsilon = \sigma\varepsilon \qquad (1.8)$$

where $\sigma = ne\mu$ is the conductivity of the material in siemens per metre. The reciprocal of σ is the resistivity in ohm-metres. Equation (1.8) is, in fact, Ohm's law on a per-unit basis, since it states that the current density is directly proportional to the voltage gradient. For a conductor of length ℓ and cross-sectional area A

$$I = JA = \sigma A\varepsilon = \sigma A.\frac{V}{\ell}$$

where $\sigma A/\ell$ is the conductance in siemens. Alternatively

$$R = \frac{V}{I} = \frac{1}{G} = \frac{1}{\sigma}.\frac{\ell}{A} = \rho\frac{\ell}{A}$$

a relationship from elementary electrical theory.
In semiconductors conduction is due to both holes and electrons carrying opposite charges and drifting in opposite directions under the influence of the applied field. If n and p are the concentrations of electrons and holes respectively and μ_n and μ_p their respective mobilities, equation (1.8) becomes

$$J = (nu_n + pu_p)e\varepsilon = \sigma\varepsilon$$

and the conductivity is

$$\sigma = (n\mu_n + p\mu_p)e$$

We have already noted that the mobilities of holes and electrons are not identical in a given situation. This is because the conduction of electrons in the conduction band differs from the process of conduction by holes in the valence band. At very low temperatures we may consider the conduction band to be empty and the top valence band filled. In this situation there can be no movement of carriers in either band. If now a small amount of thermal energy causes a number of electrons to move from the valence band into the conduction band, both these electrons and the holes consequently appearing in the valence band can contribute as we have seen to the flow of current. The small number of electrons in the conduction band can move freely but the movement of the holes in the valence band can be accomplished only when electrons, having vacated holes from elsewhere, move into them. The holes, therefore have less mobility than the electrons since they occupy fixed positions within the crystal structure and cannot occupy the intermediate positions as the electrons in the conduction band are able.

Some worked examples at this stage will illustrate the foregoing notes.

Example 1.1   Given that a copper conductor contains about 8.5×10^{28} atoms/m^3, obtain a figure for the average drift velocity of electrons in a wire having a cross sectional area of 1 mm^2 and carrying a current of 5 A.

Assuming that there is one free electron per atom and using equation (1.8) we have

$$u = \frac{J}{ne} = \frac{I}{Ane}$$

$$= \frac{5}{10^{-6} \times 8.5 \times 10^{28} \times 1.6 \times 10^{-19}}$$

$$= 3.6 \times 10^{-4} \text{ m/s}$$

This example illustrates the very slow drift velocity of the carriers in a good conductor.

Example 1.2   Given that a silicon crystal contains about 5×10^{28} atoms/m^3 and that in intrinsic material at room temperature the density of free electrons (and holes) is about 1.5×10^{16}/m^3, obtain an estimation of the intrinsic conductivity of the material. What will be the effect on the conductivity of the material if it is doped with one trivalent impurity atom per 10^9 silicon atoms? Comment.

The number of silicon atoms per electron-hole pair is

$$\frac{N_a}{n_i} = \frac{5 \times 10^{28}}{1.5 \times 10^{16}} = 3.3 \times 10^{12}$$

As we are dealing with intrinsic silicon, $n = p = n_i$ and the intrinsic conductivity will be

$$= (\mu_n + \mu_p) n_i e$$

$$= (0.135 + 0.05) \times 1.5 \times 10^{16} \times 1.6 \times 10^{-19}$$

$$= 4.44 \times 10^{-4} \ S/m$$

As there are 5×10^{28} silicon atoms/m^3, the concentration of acceptor impurity atoms is

$$N_a = 5 \times 10^{28} \times 10^{-9} = 5 \times 10^{19}$$

The corresponding intrinsic density is $p_i = n_i = 1.5 \times 10^{16}/m^3$. Therefore there are $(5 \times 10^{19}) \div (1.5 \times 10^{16}) = 3.3 \times 10^3$ as many holes. The new density of electrons is

$$n_p = \frac{n_i^2}{p_p} \approx \frac{n_i^2}{N_a} = \frac{(1.5 \times 10^{16})^2}{5 \times 10^{19}}$$

$$= 4.5 \times 10^{12}/m^3$$

At this point we pause to take stock. From this last result we see that the electron density has been reduced by the factor

$$\frac{1.5 \times 10^{16}}{4.5 \times 10^{12}} = 3.3 \times 10^3$$

which corresponds to the increase in holes. The law of mass action has been illustrated.

The new conductivity will be

$$= (n\mu_n + p\mu_p) e$$

$$= (4.5 \times 10^{12} \times 0.135 + 5 \times 10^{19} \times 0.05) \times 1.6 \times 10^{-19}$$

$$\approx (2.5 \times 10^{18})(1.6 \times 10^{-19})$$

$$\approx 0.4 \ S/m$$

This result should be compared with that of 4.44×10^{-4} S/m obtained for intrinsic silicon. The conductivity has been increased about

900 times by the addition of the very small amount (1 part in 10^9) of impurity.

Doping is thus a process involving very small amounts of added impurity which drastically affects conductivity but does not materially affect such things as the rate of thermal generation of hole-electron pairs or carrier mobility. These still depend on the properties of the overwhelmingly more abundant silicon (or germanium) atoms within the lattice.

1.5 THE p-n JUNCTION

If the doping of a melt is abruptly changed from excess donor doping to excess acceptor doping as a crystal is being grown, the crystal will be n-type material up to a certain plane and p-type beyond. This plane of transition is a p-n junction. Remembering that the excess donors in the n-type material and the excess acceptors in the p-type material number usually less than one atom in 10^7, it is remarkable that a place where the changes are so infinitesimal should exhibit such extraordinary physical properties as the p-n junction does. Yet such a junction forms the basic component as it were of all semiconductor devices, and a brief recapitulation at this point, leading on to further work on the junction diode is fully justified.

1.5.1 Carrier Diffusion

We assume that the junction consists of a single crystal structure with an abrupt change from p- to n-type material at the junction, and that the junction faces in contact are perfectly plane. At the instant of contact the carrier densities might be shown in Fig. 1.6(a)

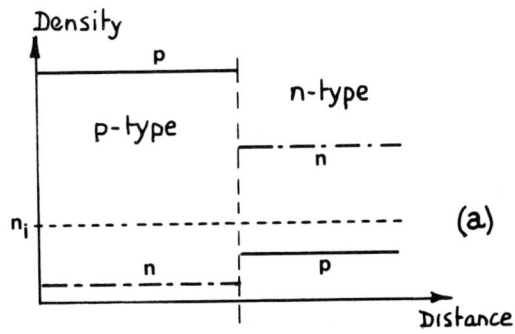

Figure 1.6(a)

15

where it is assumed that the p-type material is the more heavily doped. This situation can only be an instantaneous one; because of the high concentration of holes in the p-type material and the high concentration of electrons in the n-type material, holes migrate over into the n-material and recombine with some of the many free electrons available there, and electrons migrate over into the p-material and in turn recombine with holes. This migration is known as 'diffusion'. For any such transfer to occur, two factors must be present: there must be a concentration gradient and there must be random motion of the carriers (or particles) involved. In an analogous situation, if a small amount of dye is carefully added to a jar of water and the jar is set aside for a time without disturbance, the dye will be found to have diffused throughout the water. The thermal energy of the apparently still water provides in this case the random motion of the particles (the water molecules) and the addition of the dye provides the concentration or density gradient. No force is involved in the migration of the dye throughout the water; diffusion is a statistical redistribution which tends to equalize the concentration throughout a system. If the diffusion of the holes and electrons taking place across the p-n junction continued unchecked, the p and n densities would level out to new values uniform throughout the specimen and corresponding to the average doping. This does not happen; what does happen is that the diffusion of holes leaves uncovered immobile negative charge centres to the left of the junction and the diffusion of electrons leaves uncovered immobile positive charge centres on the right. The density distribution then settles at that shown in Fig. 1.6(b) and a space charge is established in the carrier depleted region separating the p and n materials. Hence an electric field is created internally across the junction plane. Notice that it is the movement of majority carriers across the junction that creates the internal field and that once established, this same field restrains the majority carriers on each side from diffusing without limit into the other side. Minority carriers can, however, drift across the junction, and we shall return to this aspect in a very little while. The conduction mechanism and the movement of the charges are illustrated in Fig. 1.7.

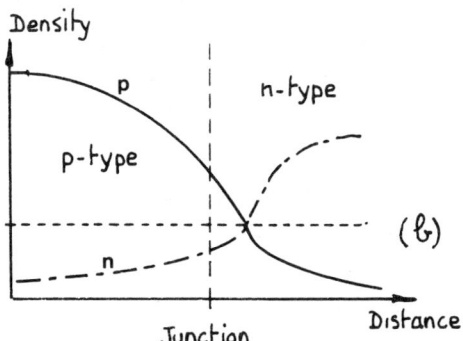

Figure 1.6(b)

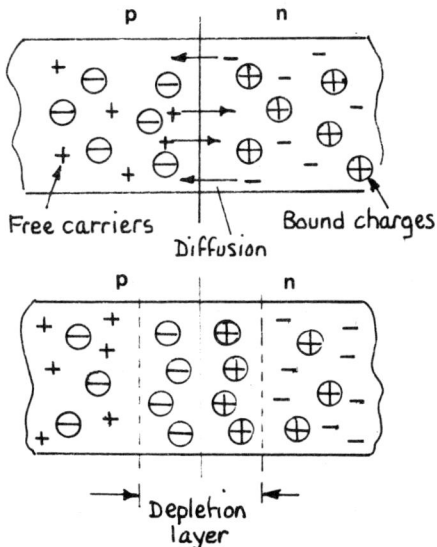

Figure 1.7

1.5.2 Potential hill

The way in which the internal field is derived is shown in Fig. 1.8. The total space charge consists of three components:

(1) A positive space charge, $+e_p$, due to the holes on the left of the junction.

(2) A negative space charge, $-e_n$, due to the electrons on the right of the junction.

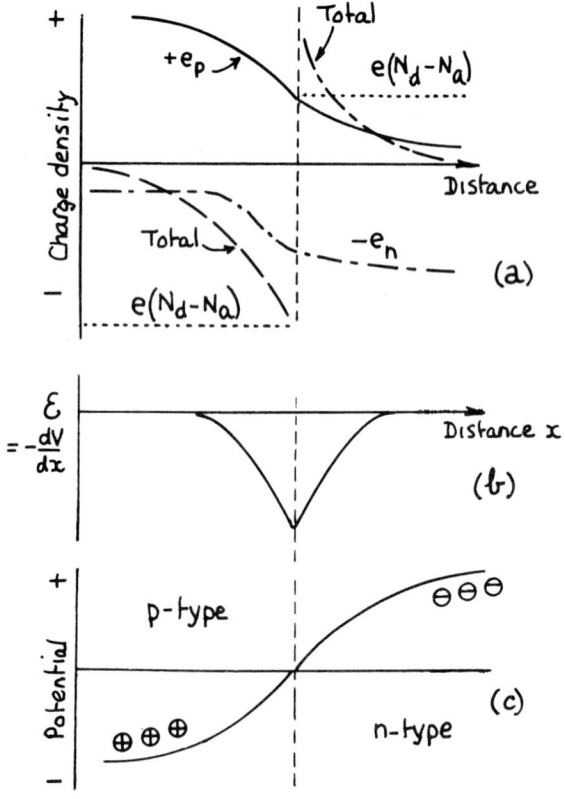

Figure 1.8

(3)   A space charge $e(N_d - N_a)$ due to the uncovered bound charges at the impurity centres.

The carrier densities shown in Fig. 1.6 must be such as to produce charge neutrality everywhere. When diffusion decreases p and increases n to the left of the junction, a net negative space charge builds up on this side. Likewise, when n decreases and p increases to the right of the junction, a positive space charge builds up there. These charge distributions are shown in Fig. 1.8(a) and correspond to the density distributions shown in Fig. 1.6(b). The electric field is the integral with respect to distance of the space charge and is shwon in Fig. 1.8(b). It is shown as negative since $\varepsilon = -dV/dx$ and is directed to the left. Under open-circuit conditions, the net current is zero and the field just balances the

diffusion everywhere. The tendency of majority electrons and holes
to diffuse from regions of high concentration into regions of low
concentration is everywhere offset by the superimposed drift of
minority carriers across the junction. The height of the potential
hill shown in Fig. 1.8(c) and its magnitude V_o is settled by the
necessity for equilibrium; for no flow of carriers the diffusion component must just equal the drift component.

With low doping levels on both sides, the value of V_o (the
'contact potential') is small. With very high dopings it can approach the energy gap of the material. Typically it is about 0.2 V
for germanium and 0.6 V for silicon. The potential hill of Fig. 1.8
(c) constitutes a barrier which holes on the p-side must climb to get
over to the n-side. We may think of the holes as small ball bearings
on the floor at the foot of a ramp, and bouncing around with various
kinetic energies. Occasionally one of the balls gets sufficient
energy to roll up the ramp and arrive at the top. Likewise, we can
imagine the electrons as buoyant balloons under the ramp at the top.
Now and again one of these balloons gets just enough energy to slip
down the underside of the ramp to the bottom, in spite of its opposing
buoyancy. There is thus a forward component of current, I_e, due to
the more energetic of the majority carriers forcing their way across
the junction in spite of the opposing field.
carriers getting across the junction is a function of the height of
the hill; only those majority carriers having kinetic energies
greater than eV_o make the trip. If equilibrium is upset by any
change in the contact potential from V_o to $(V_o \pm V)$ then the probability of majority carriers possessing sufficient energy to pass the
barrier is similarly changed. Hence the flow of carriers across the
junction depends upon the value of $(V_o \pm V)$.

Suppose the junction potential is reduced by the application of
an external voltage V_1 as shown in Fig. 1.9. The effective junction
potential is now $(V_o - V_1)$ and the probability of the majority
carriers having sufficient energy to cross the barrier is greatly
increased. Holes from the p-material cross the junction and enter
the n region where they represent minority carriers. The electrons
move in the opposite direction. The junction is now said to be
forward biased and a net injection current flows, sustained by the

19

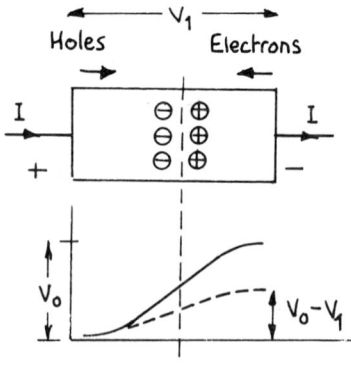

Figure 1.9

continuous generation of majority carriers and recombination of the
injected minority carriers. In the external circuit wires the only
carriers are, of course, electrons.

Suppose now the junction potential is raised by the application
of an external voltage which adds to the internal potential, see
Fig. 1.10. Since the effective barrier potential is now $(V_o + V_1)$

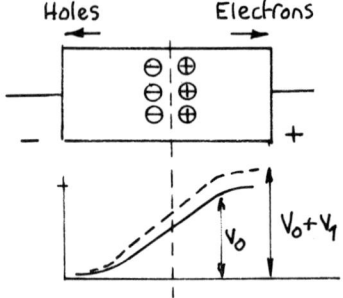

Figure 1.10

the probability of majority carriers possessing sufficient energy to
cross the junction is greatly reduced. The net migration of majority
carriers and hence the minority injection current is reduced to zero.
The junction is now said to be reverse biased.

Although only a fraction of a volt of reverse bias is enough to
prevent the diffusion of majority carriers, a small reverse current
does however flow. This is due to the thermally generated minority

carriers appearing in the vicinity of the junction being constrained
to drift in the direction of the field. At the junction itself,
minority holes in the n-type material roll 'down' the potential hill
and minority electrons in the p-material roll 'up'. No additional
energy is needed by them to move this way; in fact, they gain energy
by their change in levels. This reverse current flow is known as the
reverse saturation or leakage current, I_s. I_s depends upon the rate
of thermal generation only and hence upon temperature. The height of
the potential hill does not affect it, hence it is independent of
the applied voltage V_1. For very low values of reverse bias, I_s
reaches its maximum value and thereafter remains constant; hence the
term 'saturation' current.

1.6 THE DIODE EQUATION

We have seen that the total potential across the p-n diode junction
is the equilibrium or contact potential V_o less any external forward
bias voltage V. Thus we can expect I_e to increase with forward bias.
It can be shown that the probability of a carrier possessing sufficient kinetic energy to cross the potential barrier is expressed in
terms of $\exp eV_o/kT$, hence we may write

$$I_e = I_o . \exp \frac{eV}{kT} \qquad (1.9)$$

where V is the applied voltage (positive for forward bias), T is
the absolute temperature and k is a constant (Boltsmann's constant =
1.38×10^{-23} J/°K). We shall introduce I_o a little later on.

In addition to the above forward component of current, there is
the reverse saturation component I_s due to the minority carriers from
both sides which diffuse to the vicinity of the junction and get
swept across by the field. Since the minority carrier densities outside of the space charge region are independent of applied voltage,
I_s is, as we have noted, constant and negative. Thus we have two
components of current, one voltage dependent, the other not. The
total diode current is the algebraic sum of the injection current
I_e and the saturation current I_s, so

$$I = I_e - I_s = I_o . \exp \frac{eV}{kT} - I_s$$

With no applied voltage, V = 0, the total current I must be zero, so $I_o = I_s$ and we have

$$I = I_s \left[\exp \frac{eV}{kT} - 1 \right] \tag{1.10}$$

This is the general junction diode equation. It is reasonably accurate for all germanium diodes and silicon diodes at high currents.

For reverse bias, V is negative and the first term of (1.10) quickly becomes negligible. Thus the reverse current saturates to a constant value I_s. At room temperature, say T = 20°C = 293 K, e/kT = 38.6 (say 40) and a convenient form of the equation becomes

$$I = I_s (\exp 40\ V - 1) \tag{1.11}$$

or, changing bases

$$= I_s (10^{17\ V} - 1) \tag{1.12}$$

Clearly, under even a small forward bias, exp 40 V >> 1 and from then on the current increases exponentially by a factor e ($\simeq$ 2.72) with every 25 mV of forward bias, or by a factor of 10 about every 60 mV.

Example 1.3   At room temperature the reverse saturation current of a germanium diode is 80 μA at an applied voltage of -1 V. Estimate the diode current for a voltage of -0.1 V and +0.1 V under the same ambient condition.

Here I_s = 80 μA. For V = -0.1 V, $e^{40\ V} = e^{-4}$
$$\simeq 0.02$$

Therefore $I \simeq -I_s \times 0.98 \simeq -78.4$ μA.

For V = +0.1 V, $e^{40\ V} = e^4 \simeq 55$

Therefore $I = (80 \times 10^{-6})(55^{-1}) \simeq 4.3$ mA

Figure 1.11 shows the voltage-current characteristics of both silicon and germanium diodes.

If we differentiate equation (1.10) we have

$$\frac{dI}{dV} = \frac{e}{kT} I_s \cdot \exp \frac{eV}{kT} = \frac{e}{kT}(I + I_s)$$

So the dynamic resistance of the diode (the gradient of the

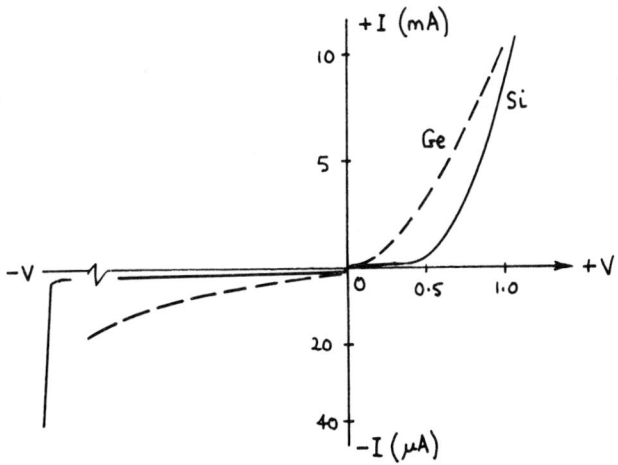

Figure 1.11

characteristic)

$$R = \frac{dV}{dI} = \frac{e}{kT}(I + I_s)^{-1}$$

Putting in the appropriate values for e, k and T and expressing the current in milliamperes, reduces this to the simple form

$$R \approx \frac{25}{I + I_s}$$

For forward currents of 0.1 mA or more, I_s can normally be neglected.

Equation (1.10) shows that the diode current is a function of the temperature, which turns up in the denominator of the exponent. However, the variation of the saturation current with temperature greatly exceeds the variation in the exponential term. As stated earlier, I_s depends upon the minority carrier densities in the p- and n-type materials. These densities, as we have seen, are

$$n_p \approx \frac{n_i^2}{N_a} \quad \text{and} \quad p_n \approx \frac{n_i^2}{N_d}$$

I_s therefore varies as n_i^2 and since n_i varies exponentially with temperature, so does I_s. By some mathematical reasoning it is found that at 300 K (room temperature) the reverse saturation current should double about every 10°C for germanium and every 5°C for silicon. This is found to be the case for germanium. In silicon at ordinary temperatures the ideal saturation current is so low that the reverse current is dominated by surface leakages and other effects which vary less rapidly with temperature.

1.7   FABRICATED RESISTORS

We have noted how the conductivity of semiconductor materials can be dramatically changed by the introduction of small and controlled amounts of impurity. This property enables a complete circuit containing not only semiconductor junctions (and hence diodes and transistors), but resistors, conductors, insulators and capacitors to be fabricated on a single wafer or chip of material by selective control of areas and concentrations of doping. Such circuits are known as 'integrated' or 'monolithic' circuits and these increasingly form the basic building blocks of modern electronic systems.

Example 1.4   A bar of intrinsic germanium measuring 1 mm square by 5 mm in length has about 2.5×10^{19} free electrons/m³ at room temperature. What would be the resistance of this bar at room temperature?

Conductivity $\sigma = e(n\mu_n + p\mu_p)$

Using the mobility figures given in Table 1.1, where $\mu_p = 0.19$, $\mu_n = 0.4$, we have

$\sigma = 1.6 \times 10^{-19}(2.5 \times 10^{19} \times 0.4 \times 2.5 \times 10^{19} \times 0.19)$

$ = 1.6 \times 2.5 \times (0.4 + 0.19)$

$ = 2.36$ S/m

The resistivity $\quad \rho = \dfrac{1}{\sigma} = \dfrac{1}{2.36} = 0.424$ Ω-m

Therefore $\quad R = \dfrac{\rho l}{A} = \dfrac{0.424 \times 5 \times 10^{-3}}{0.001^2}$

$ = 2120\,\Omega.$

Some problems on the contents of this chapter now follow.

1.8   A LABORATORY EXPERIMENT

You will need a silicon diode and a germanium diode. This latter can
be a point contact type or, if you want to work with a junction, get
a p-n-p transistor (one of the older OC series such as the OC71 is
suitable) and use the base-emitter diode for the experiment. Either
ignore the collector lead or join it to the base.

Set up the circuit shown in Fig. 1.12(a) and with the diodes inserted in turn, increase the voltage across the diode in steps of
0.1 V up to about 1 V and thereafter in steps of 0.2 V until the circuit current is about 10-15 mA. Record the voltage and current
values.

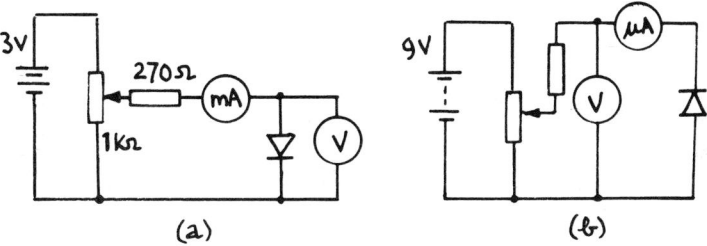

Figure 1.12

With the circuit wired now to Fig. 1.12(b), measure the reverse
current of each diode on the microammeter for reverse potentials up
to about 10 V. It is quite probable that the reverse current through
the silicon diode will be undetectable, but for the germanium diode
it will rapidly rise beyond a reverse voltage of a few volts. Make
a note of the room temperature throughout. Record the voltage and
current values.

Plot voltage horizontally against current vertically for both
diodes after the fashion of Fig. 1.11.

Now use the equation (1.10) in the text to determine the value
of e/k. Hint: equation (1.10) can be approximated as $I = I_s . \exp\frac{eV}{kT}$,
then taking logs

$$\ln I = \ln I_s + \frac{eV}{kT}$$

If ln I is now plotted against V, a straight line graph will result.
You can use log-linear paper here. From the gradient of this graph,

for a known value of temperature T, the ratio e/k can be determined. Given that k = 1.38×10^{-23} J/$^\circ$K or 8.62×10^{-5} eV/$^\circ$K, estimate the charge on the electron.

PROBLEMS 1

1. Define the terms: conduction band, valence band, covalent bond, mobility, intrinsic, recombination.

2. How does the speed of a signal (say a voice) transmitted along a conducting wire compare with the drift velocity of the electrons?

3. Why is silicon preferred to germanium in semiconductor devices nowadays?

4. A semiconductor is doped with donor atoms. What effect does this have on the electron and hole densities?

5. Explain qualitatively why the resistance of pure metals increases with temperature while the resistance of semiconductors decreases.

6. State a form of the mass-action law. Would it be correct to express it as saying that the rate of recombination depends upon the number of reacting elements present?

7. Show that the following approximation is justified:

$$a = e(n\mu_n + p\mu_p) \simeq N_a \mu_p e$$

8. When drifting under the influence of an electric field of intensity 315 V/m, the velocity of electrons in a piece of n-type material is about 120 m/s. What is the electron mobility? What are the units of mobility?

9. Does the term n_i^2 vary with temperature? If so, in what way does it vary?

10. If a semiconductor is doped with equal concentrations of donor and acceptor atoms, how does the material behave electrically?

11. A silicon crystal at room temperature contains about 1.5×10^{16} free electrons (and holes) per m^3. If the crystal contains 5×10^{28}

atoms/m^3 how many carriers are there per atom in the undoped material?

12. The concentration of free electrons in intrinsic germanium at room temperature is about 2.5×10^{19}/m^3. What would be the resistance of a bar of germanium having a cross-sectional area of 2 mm^2 and of length 10 mm?

13. A bar of intrinsic silicon having the same dimensions as that of the previous example has about 1.5×10^{16} free electrons/m^3 at room temperature. Show that the resistance of the bar is about 11 MΩ.

14. The reverse saturation current for a certain diode is 10 nA. Sketch a current-voltage characteristic for this diode by calculating current magnitudes for forward bias voltages from 0.1 V to 0.6 V (in steps of 0.1 V), and for reverse bias voltages from -0.1 to -1 V.

15. At room temperature a silicon crystal contains about 5×10^{28} atoms/m^3. If every 10-millionth silicon atom is replaced by a p-type impurity atom, show that the resistivity will be decreased by a factor of about 9×10^4.

16. In intrinsic germanium, hole and electron mobilities are respectively 0.19 and 0.4 m^2/V-s. If the measured conductivity of a specimen is 2.5 S/m calculate the density of hole-electron pairs.

2 Properties of semiconductor devices

2.1   TYPES AND CONFIGURATIONS

We shall be concerned in this chapter with the fundamental operation
and fabrication of discrete transistors: the bipolar junction transistor in which both types of current carrier (electrons and holes) are
concerned, and the unipolar or field-effect transistor (FET) in which
only one type of carrier (electrons OR holes) are involved.

   We recall that the transistor, being a three-terminal device, may
be connected in any one of three ways or basic circuit configurations:
(a) common-base, (b) common-emitter, and (c) common-collector mode
for the bipolar transistor; and (a) common-gate, (b) common-source,
and (c) common-drain mode for the FET. In all three configurations
one electrode is common to the input and output terminal pairs and
this common electrode is, in general, treated as being at earth or
chassis potential. The term 'earthed' or 'grounded' is sometimes
used in place of the term 'common' to describe the particular mode
of connection so that the common-base configuration may, for example,
be referred to as the earthed- or grounded-base mode. For the
realisation of power gain, the base must always be associated with
one of the input terminals. There are no practical applications in
which any of the three basic configurations are used in reverse.
The configurations are illustrated in Fig. 2.1.

2.2   THE JUNCTION TRANSISTOR

Essentially, the junction transistor consists of a pair of p-n
junctions connected back to back. Like the p-n diode, it is formed
of a single crystal with the added impurities distributed so as to

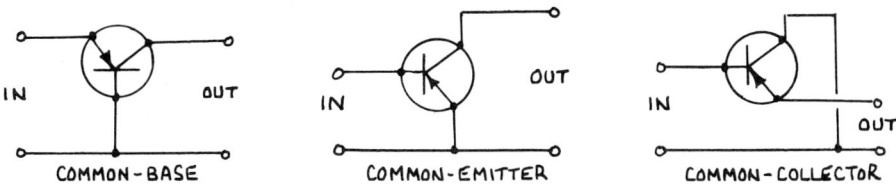

Figure 2.1

produce two junctions; either an n-type layer sandwiched between two
p-type layers, or a p-type layer sandwiched between two n-type layers.
The two types are known as p-n-p and n-p-n respectively. Their
operation is essentially similar except for the polarity of the
applied voltages and, because holes are the majority carriers in the
p-n-p form and electrons in the n-p-n form, the direction of current
flow. The layers, taken in order, are known as the collector, base
and emitter respectively; hence the transistor is a three terminal
device. Our discussions will, in the main, concern the n-p-n transistor.

2.2.1   General fabrication

There are a number of ways of forming the three necessary layers of
a junction transistor but only the 'planar' method will be mentioned
here. This method is the basic technology in the fabrication of integrated monolithic circuits and we can usefully introduce it at this
stage as it is applied to the production of discrete transistors.

   As the name implies, planar fabrication is carried out on and
parallel to the surface of a wafer of semiconductor material. A
property of silicon is that if a base (or substrate) wafer is subjected to an oxidizing atmosphere, a layer of silicon oxide is
formed on the surface of the wafer and this acts as a barrier to the
diffusion of certain elements including phosphorus and boron (n-type
and p-type impurities respectively) into the silicon. The process
of planar construction begins therefore with a doped (say, n-type)
wafer which receives a uniform protective layer of silicon oxide;
see Fig. 2.2(a). If now a selected area of this oxide layer is
etched away, the silicon substrate will be exposed and impurity
atoms can be diffused into this region without contaminating the

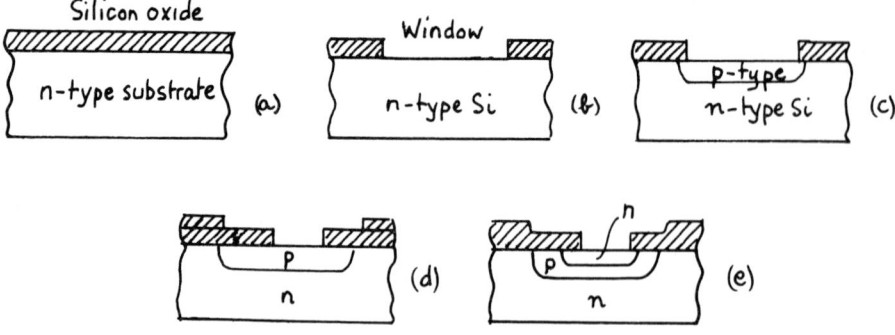

Figure 2.2

rest of the slice. The removal of the oxide layer at the required
place is achieved by a photo-engraving process, in general a procedure rather similar to that used in the manufacture of ordinary
printed boards but on a much smaller scale and with much greater
accuracy. A layer of photo-resist lacquer is sprayed on the oxide
layer; this lacquer is sensitive to ultra-violet radiation and when
exposed to such radiation is hardened to resist attack by acids and
solvents. After the oxide layer has been thus exposed through a
suitable photographic mask, that area which has not been subjected
to the light is removed by solvent and a 'window' is formed in the
oxide layer that remains, see Fig. 2.2(b). The layer is now placed
in a p-type atmosphere maintained at an elevated temperature with
the result that the p-type impurity atoms diffuse into the silicon
through the window opening but do not diffuse into the surrounding
oxide protected area. A plane or layer of p-type material is then
established in the substrate as shown in Fig. 2.2(c).

A further layer of silicon oxide is now formed over the entire
substrate surface so that the original window opening is blotted
out while the original oxide layer is slightly thickened in the
other parts. The photo-resist process is repeated to remove the
oxide layer in a select-area and a second window is created as shown
in Fig. 2.2(d). Again the wafer is placed in an impurity atmosphere
(this time of n-type) and a layer of n-type silicon is diffused into
the previously established p-type region as shown in diagram (e).
Finally, a further layer of oxide is formed over the wafer and
selectively removed from those areas where ohmic contacts are to be

30

made to the emitter, base and collector layers. It should be noted that the configuration reached in Fig. 2.2(c) could be used as a diode.

Such transistors are not made individually. A slice of silicon (in its turn cut from a large crystal of silicon) having a diameter of possibly 5 cm constitutes the substrate wafer and many hundreds of discrete transistors are formed at the same time. Each transistor element, probably only a millimetre or so square, is finally broken away from the parent slice and individually packaged in a suitable casing for protection against mechanical and environmental damage.

The planar form of construction has two important advantages over other methods: firstly, all dimensions are determined by a photographic process which leads to tight dimensional control and hence to tight electrical specification; secondly, the semiconductor surfaces are all protected by silicon oxide as the junction is made and this makes the effect of oxygen and water vapour less likely to affect the operation of the transistor if the housing is not hermetically sealed against the atmosphere.

2.3   FUNDAMENTAL TRANSISTOR ACTION

Consider a thin slice to be made through the three layers of an n-p-n transistor as shown in Fig. 2.3. The application of a forward bias voltage V_{BE} to the base-emitter junction causes the electrons in the n-material to be repelled from the emitter region. The emitter therefore acts as a source of electrons which flow into the base. In the base region the electrons drift towards the collector by the process of diffusion and are swept up the potential hill into the collector which is biased positively with respect to the base. Here they are effectively removed by the external battery (V_{CB}), leaving behind a supply of holes available for recombination.

Because the base is p-type, electrons exist there only as minority carriers. The electrons arriving at the collector are derived almost entirely from the emitter by way of diffusion through the base. A very small proportion, typically less than 1% of the electrons recombine with holes in the base and this loss of charge is made good by the flow of a small base current I_B. Variation of the base current varies the voltage across the base-emitter junction

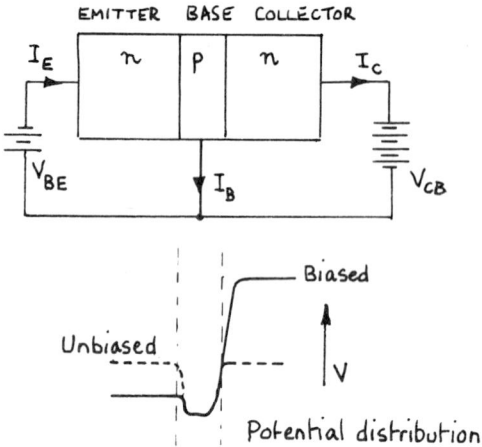

Figure 2.3

and so controls the collector current I_C, thus enabling current
amplification to be achieved.

For a p-n-p transistor the majority carriers in the emitter are
holes. The emitter acts as a source of holes which flow into the
base when this is negatively biased with respect to the emitter.
The collector, which is now biased negatively with respect to the
base, absorbs the majority of the emitted holes, the remainder re-
combining with majority electrons in the base region.

From a direct voltage viewpoint the transistor, as mentioned
earlier, can be looked on as a pair of p-n diodes connected back to
back; the emitter-base diode is then normally forward-biased and the
base-collector diode is reverse-biased. 'Make the p-material positive for forward biasing' is a useful rule to keep in mind when considering diodes, and conversely for reverse biasing. The representation of a transistor as two discrete diodes does not however describe
its active performance in any manner.

2.3.1   D.C. Behaviour
There is clearly a definite relationship between the currents flowing
at the terminals of a transistor since both I_B and I_C are derived
from the emitter current I_E. Hence

$$I_E = I_C + I_B$$

where, in practice

$$I_E \simeq I_C$$

For efficient operation as much as possible of the emitter current should reach the collector. We consider the ratio I_C/I_E. The current from the forward biased emitter junction, I_E, consists (for an n-p-n transistor) of two components:

(a) electrons diffusing from emitter to base,

(b) holes diffusing from base to emitter as reverse saturation or leakage current.

In a diode proper, the relative magnitudes of these components are, in general, of minor importance. In the transistor however, only the 'electron' component has any chance of getting through to the collector. The holes are travelling in the wrong direction for a start, and in any event they would be repelled by the positive field at the collector. Thus the useful fraction of the input current is represented by the ratio

$$\gamma = \frac{\text{electron component of the emitter current}}{\text{total emitter current}}$$

where γ is known as the emitter efficiency. To maintain γ close to unity, the electron component of I_E should be large and this requires that the emitter should be more heavily doped than the base.

Having got the maximum proportion of I_E to enter the base as electrons, it now becomes necessary to ensure that as many of these as possible actually reach the collector. If the base region is very narrow, the potential difference acting across the opposite faces of the base will be small so that once in the base the electrons will move only by the process of diffusion. At the emitter side of the base the electron density will be high since this is their area of injection. At the collector side of the base the density will be low since the collector is gathering them up as fast as they arrive. Since diffusion is a process which tends to equalise concentration throughout a system, the narrower the base region the better. The electrons do not then spend too much time in the base region and so avoid elimination by recombination with base majority holes. The ratio

$$\beta = \frac{\text{electron current reaching the collector}}{\text{electron component of emitter current}}$$

is known as the base transport factor.

The overall efficiency of current transfer between emitter and collector is clearly given by the product $\gamma\beta$. This is

$$\frac{\text{electron current reaching the collector}}{\text{total emitter current}} = \frac{I_C}{I_E}$$

We recognise this ratio as α_B (h_{FB}), the static current gain or forward current transfer ratio in common-base configuration. Because only d.c. conditions are involved it is usual to speak of α_B as being the low-frequency, short-circuit current gain of the transistor. This ratio must, of course, be less than unity.

2.3.2   Leakage current

Let a transistor connected in common-base configuration have its emitter lead disconnected as shown in Fig. 2.4. Clearly I_E will now be zero and it might be suggested that I_C will be zero also. The collector circuit is, however, still completed through the reverse-biased base-collector junction and this diode must pass a reverse saturation current by virtue of the thermally generated minority carriers (holes) crossing from collector to base. This current flows in the external circuit in the SAME sense as that due to collected electrons. It is denoted by I_{CBO}, indicating a collector to base flow with the emitter open-circuited. This current will continue

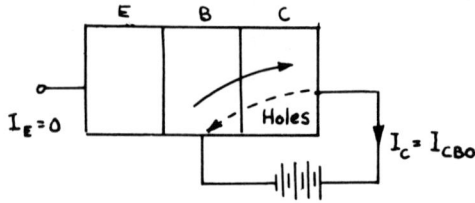

Figure 2.4

to flow more or less unchanged when the emitter is reconnected and the main emitter current is superimposed. Hence

$$I_C = \alpha_B I_E + I_{CBO} \tag{2.1}$$

Here the term $\alpha_B I_E$ represents the USEFUL component of current because the amplifying properties of the transistor depend upon it; the second term is unwanted but is nevertheless always present.

There may be considerable variation in I_{CBO} between otherwise identical devices and it is temperature sensitive, particularly in the case of germanium transistors where, at room temperature, it may be of the order of several microamperes. In silicon devices it is normally of negligible magnitude. A slight increase in I_{CBO} occurs with an increase in collector voltage, but at any particular temperature level it may be considered constant.
Since

$$I_B = I_E - I_C$$
$$I_B = I_E(1 - \alpha_B) - I_{CBO} \tag{2.2}$$

from equation (2.1) above.

A summary of the currents acting in the various regions of an n-p-n transistor might usefully be made at this point:

(a) I_E is primarily electrons moving from emitter to base but also some holes moving from base to emitter.

(b) I_B is primarily holes that recombine with the main flow of electrons diffusing through the base region but also I_{CBO} as noted above plus holes moving from base to emitter. Externally, I_B appears as an electron flow LEAVING the base.

(c) I_C is primarily electrons derived from the emitter but also includes I_{CBO}, holes moving from collector to base plus some electrons as minority carriers moving from base to collector just as in a diode. From equation (1.10) earlier, the collector cut-off current is given by

$$I_{CBO}(\exp \frac{eV}{kT} - 1) \tag{2.3}$$

which approximates to I_{CBO} when V (negative) exceeds a fraction of a volt.

All current flow in wires external to the transistor is electron movement. Study Fig. 2.5 very carefully; it shows the distribution of the component currents for an n-p-n transistor.

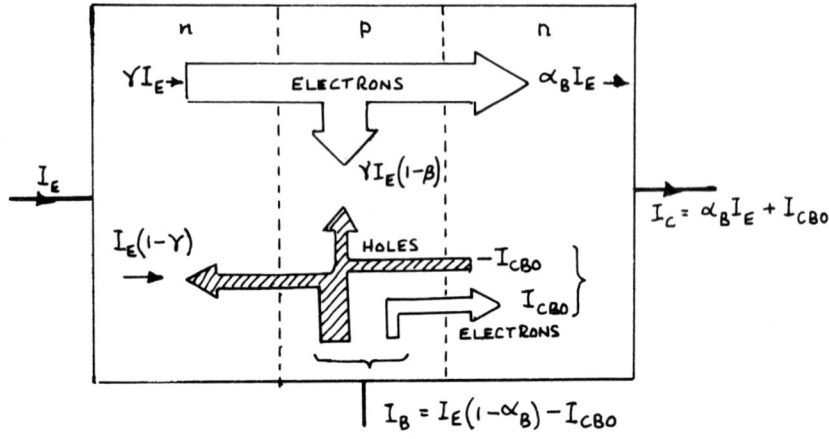

Figure 2.5

PROBLEMS

2.1   If I_C = 2.75 mA and I_E = 2.71 mA, what is I_B? If, in addition, I_{CBO} = 5 μA, what is α_B? If the base of this transistor is disconnected, what will be I_C?

2.2   Sketch, after the manner of Fig. 2.5, the current distributions and directions of flow for a p-n-p transistor.

2.4   CONFIGURATION CHARACTERISTICS

The principles upon which a transistor operates are not affected by the circuit configuration employed. Inside the transistor the physical actions already discussed are unchanged whichever mode we happen to be using. We shall examine briefly the main characteristics of the common-base and common-emitter configurations, leaving the common-collector (emitter-follower) for a later analysis.

2.4.1   Common-base characteristics

The common-base connection is shown in Fig. 2.6. The input characteristics of this configuration can be deduced from a knowledge of diode characteristics. Since the emitter-base junction is a forward-

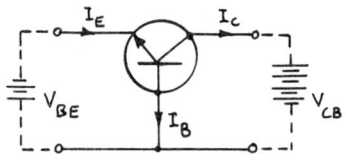

Figure 2.6

biased diode, the characteristics of such a diode are essentially
the input characteristics of the transistor. These are shown in
Fig. 2.7(a) where the effect of variations in the collectorbase voltage, V_{CB}, is indicated. With the emitter open-circuited, $I_E = 0$ and
the base-collector junction is a reverse-biased diode. For $I_E = 0$,
$I_C = I_{CBO}$ and the collector (or output characteristic) is identical
with the reverse-bias characteristic of the equivalent diode. For
I_E finite, say $I_E = 1$ mA, the collector current increases by an
amount $\alpha_B I_E \simeq 1$ mA, and the appropriate curve is shown in Fig. 2.7(b).
Notice that the collector cut-off current is I_{CBO} when V_{CB} is reversed
by a fraction of a volt. The current gain of the common-base configuration is always less than unity since I_E is always greater than I_C.

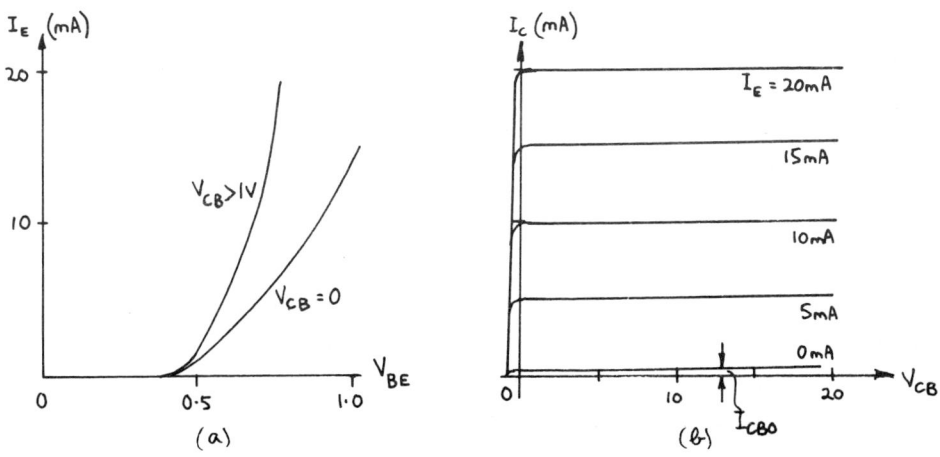

Figure 2.7

Although the lines appear to be parallel and horizontal, indicating that α_B is constant, they actually slope slightly upwards since
α_B increases with increases in V_{CB}. As V_{CB} varies, the width of the

collector-base depletion layer also varies and since this layer includes part of the base wafer, the effective base width also varies.
This is an effect known as 'base width modulation'. As V_{CB} increases
the depletion layer widens and the base width decreases; electrons
therefore have less distance to diffuse before reaching the collector
and fewer are lost in recombination with holes in the base. Hence
the base transport factor β is increased and so is the current gain
α_B.

Base width modulation also affects the input characteristic by
making this dependent also upon V_{CB}. When V_{CB} is reduced the base
transport factor decreases and the effective base width increases.
The concentration of electrons in the base will not fall off quite
so rapidly under this circumstance and the injection of carriers from
the emitter into the base is restricted by the negative charge facing
them. Thus for a given I_E, V_{BE} has to increase as V_{CB} decreases,
Fig. 2.7(a) shows the effect on the input characteristic.

2.4.2   Common-emitter characteristics

Consider now the common-emitter connection of Fig. 2.8 where the base
current I_B is treated as the input. As before

$$I_C = \alpha_B I_E + I_{CBO}$$

so that

$$I_C = \alpha_B (I_C + I_B) + I_{CBO}$$

or

$$I_C (1 - \alpha_B) = \alpha_B I_B + I_{CBO}$$

Therefore

$$I_C = \frac{\alpha_B}{1 - \alpha_B} I_B + \frac{1}{1 - \alpha_B} I_{CBO}$$

But we recognize $\alpha_B/(1 - \alpha_B)$ as being α_E, the static current gain in
common-emitter. Hence

$$I_C = \alpha_E I_B + \frac{1}{1 - \alpha_B} I_{CBO} \tag{2.4}$$

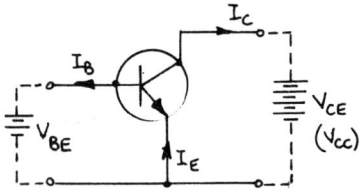

Figure 2.8

So for $I_B = 0$, that is, with the base open-circuited

$$I_C = \frac{1}{1 - \alpha_B} I_{CBO}$$

This term must then represent the leakage current in common-emitter mode and is denoted I_{CEO}, the current flowing between collector and emitter with the base open-circuited. This current is clearly I_{CEO} multiplied by the static current amplification of the transistor since $1/(1 - \alpha_B) = \alpha_E + 1 \simeq \alpha_E$. This can typically lie in the range 50 to 500.

Typical common-emitter input and output characteristics are shown in Fig. 2.9(a) and (b). The input current I_B is small and for a collector-emitter voltage V_{CE} greater than about 0.5 V depends only upon the emitter-base voltage V_{EB}. With the base open-circuited, $I_B = 0$ and the collector cut-off current is I_{CEO}. This current, which is practically constant has been exaggerated for clarity in Fig. 2.9(b). For I_B finite, each increment in I_B causes the

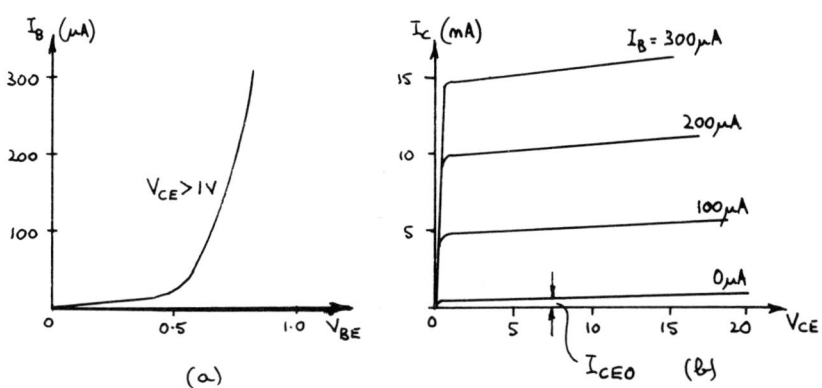

Figure 2.9

collector current to increase by $\alpha_E I_B$ and because α_E is >> 1, the current gain is large.

PROBLEMS

2.3 Consider the common-collector configuration where the base current is considered as input and the emitter current as output. Write down an equation similar in form to (2.4) above for the total emitter current. Hence, or otherwise, show that $\alpha_C = \alpha_E + 1$.

<u>Example 2.1</u>   A silicon n-p-n transistor having $\alpha_B = 0.995$ and $I_{CBO} = 1$ nA is connected as shown in Fig. 2.10. Calculate values of I_C, I_E and V_{CE}.

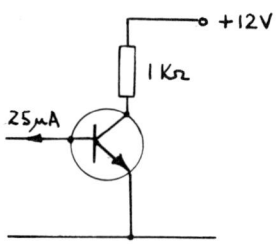

Figure 2.10

Here $\alpha_B = 0.995$, hence $\alpha_E = \dfrac{0.995}{1 - 0.995}$

$$= 199$$

The collector cut-off current I_{CEO} is then 200×1 nA = 200 nA = 0.2 μA. As expected of a silicon device, I_{CEO} is negligible relative to the collector current which is

$$I_C = (200 \times I_B) + I_{CEO}$$

$$= (200 \times 25) + 0.2 \ \mu A = 5 \ mA$$

The emitter current is $I_C + I_B = 5.025$ mA.

The collector-emitter voltage V_{CE} is equal to the supply voltage V_{CC} less the voltage drop across the collector load resistor R_L. Hence

$$V_{CE} = 12 - I_C R_L = 12 - (5 \ mA \times 1 \ k\Omega) = 7 \ V$$

40

Since $V_{CB} = V_{CE} - V_{BE}$ = 7 - 0.6 = 6.4 V, the reverse bias on the collector-base diode is -6.4 V.

2.5 THERMAL INSTABILITY

The previous example illustrated that the effect of I_{CBO} in a silicon transistor, even after amplification in the common-emitter mode, is generally negligible in comparison with the normal current at the collector. It is of interest to consider the germanium transistor in this respect. Not only is the magnitude of the leakage current in a germanium device very much larger than that in silicon, but it is much more temperature dependent, in so far as a given temperature rise leads to an increase in the leakage current which, because of its 'start' on the silicon value, is many times the corresponding increase suffered by the silicon. Typical values for α_B and I_{CBO} in a germanium transistor will be α_B = 0.99, I_{CBO} = 1 μA at 20°C, rising to some 10 μA at 50°C. For these figures, I_{CEO} will be of the order of 100 μA at 20°C and 1 mA at 50°C. Such a level may well be comparable with the useful collector current $\alpha_E I_B$. Such an increase will cause the temperature of the collector junction to rise, I_{CBO} will increase further as will I_{CEO}, and this in turn will lead to a further heating of the junction. Such a situation will lead to thermal instability and the transistor may be quickly destroyed.

The substitution of transistors having different values of α_B may also lead to this problem. For a transistor having α_B = 0.98, $1 - \alpha_B$ = 0.02 and so α_E = 49. If this transistor is now replaced by another having α_B = 0.99, $1 - \alpha_B$ = 0.01 and α_E = 99. Hence a change of 1% in the value of α_B has led to a change of 100% in the leakage current I_{CEO}. To summarise in a sentence: the importance of the effects of leakage current lies in the effect its variations have upon the stability of collector current and hence upon the stability of the operating point. We shall return to this problem later on.

Example 2.2   At room temperature, two germanium transistors A and B, have α_B values of 0.97 and 0.99 respectively. Also A has I_{CBO} = 5 μA and B has I_{CBO} = 10 μA. What are the α_E and I_{CEO} values for these transistors?

If the collector current of A is set at 1 mA, find I_B. If this value of I_B is maintained constant, what will be the collector current of B when it is substituted for A?

For transistor A: $\quad _E = \dfrac{B}{1 - \alpha_B} = \dfrac{0.97}{0.03} = 32.3$

Therefore $\quad I_{CEO} = 33.3 \times 5 \text{ }\mu A = 167 \text{ }\mu A$

For transistor B: $\alpha_E = \dfrac{0.99}{0.01} = 99$

Therefore $\quad I_{CEO} = 100 \times 10 \text{ }\mu A = 1 \text{ mA}$

For A: $\quad I_C = 32.3 I_B \times 167$

$\quad\quad I_B = 25.8 \text{ }\mu A$

For B: $\quad I_C = (99 \times 25.8) + 1000 \text{ }\mu A$

$\quad\quad\quad = 3.55 \text{ mA}$

Such a shift in I_C by transistor replacement would be intolerable. Obviously the information in this example was 'rigged' to make a point, but the end result illustrates the problem of stability.

2.6 THE FIELD-EFFECT TRANSISTOR

The field-effect transistor (FET) is the solid state counterpart to the thermionic valve and operates upon a completely different principle from the bipolar transistor. In the thermionic valve, anode current is controlled by varying the electric field established by the grid potential. In the FET current flowing along a channel of semiconductor material is controlled by the width of a depletion region established between p-n junctions situated at a point along the length of the channel. The FET depends upon only one type of charge carrier and for this reason is known as a unipolar device.

2.6.1 The junction-gate FET

In the junction-gate FET of JFET, a narrow conducting length of p- or n-type semiconductor has ohmic contacts made at each end; the length of material is known as the channel and the end connections as the 'source' and 'drain'. Between the source and the drain terminals, the channel passes between 'gates' formed from semiconductor material of opposite type to the channel. The general planar con-

struction is shown in Fig. 2.11 with a simplified representation in
Fig. 2.12. We assume here that the channel is n-type material. When
a voltage V_{DS} is established between the source and drain, a current
will flow along the channel (conventionally) from drain to source.
This drain current will clearly depend upon the value of V_{DS} and the
channel dimensions. If now the p-n junctions are reverse biased by
applying a voltage V_{GS} to the gate terminal, a depletion layer will
be formed on each side of the channel; as this depletion region will
be free of mobile carriers, the effective cross-sectional area of the
channel will be reduced. As the channel has finite conductivity
there will be a linear fall in potential along its length from drain
to source, hence the depletion layers will be asymmetric because
the reverse bias will be higher at the drain end of the channel than
at the source end. The flow of carriers (electrons in the n-channel)

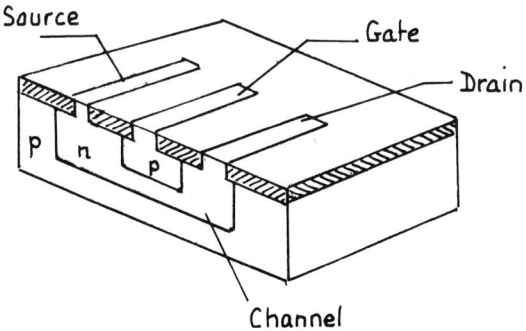

Figure 2.11

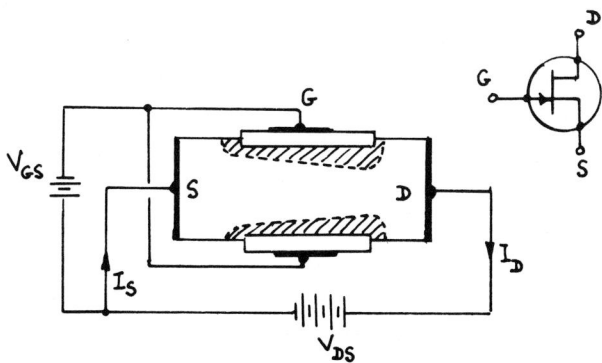

Figure 2.12

43

will now be restricted to the wedge shaped profile as Fig. 2.12 shows. Clearly the effective cross sectional area of the channel is dependent upon V_{DS} as well as V_{GS}; for a fixed value of V_{GS} (including zero voltage) an increase in V_{DS} will raise the level of the potential gradient existing along the junction faces and the depletion layers will widen. At some value of V_{DS}, for $V_{GS} = 0$, the depletion layers will just meet across the channel section. The channel is then said to be 'pinched-off' and the value of V_{DS} at which this occurs is known as the pinch-off voltage V_p.

It is important to make note of the fact that the drain current does not fall to zero when the drain voltage reaches the pinch-off value because a voltage equal to V_p still exists between the pinch-off point and the source, and the electric field along the channel causes the electrons to drift from source to drain. As V_{DS} is increased beyond pinch-off the depletion layers thicken and the additional drain voltage is effectively absorbed by the enhanced field in the extended pinch-off region. The electric field between the original pinch-off point and the source remains substantially unchanged, so that the channel current and hence the drain current remains approximately constant. This is the condition of drain current saturation. Electrons arriving at the pinch-off region find themselves faced with a positive potential and are swept through the depletion region in exactly the same way as electrons are swept from base to collector in a bipolar n-p-n transistor.

If the gate terminal is maintained negative with respect to source, pinch-off occurs at a lower value of V_{DS} and the drain current is limited to a lower saturation level. Fig. 2.13 shows a

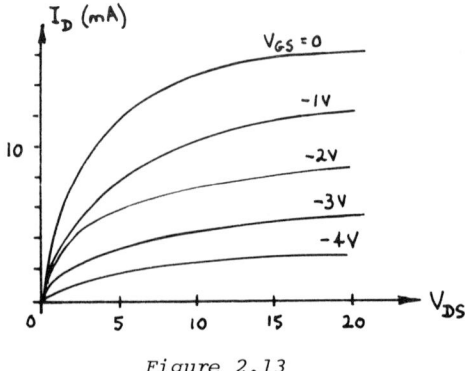

Figure 2.13

typical family of drain (or output) characteristics relating I_D to V_{DS} for constant values of V_{GS}. Below pinch-off the curves approximate to the linear Ohm's law relationship; above pinch-off in the saturation region the curves are relatively flat. This region is the normal operating range. The superficial resemblance of these characteristics to those of the bipolar transistor should be remarked.

In n-channel FET's the gate is maintained negative with respect to the source to achieve pinch-off. The polarity must be reversed (as must the polarity of V_{DS}) for a p-channel device. The sign of V_p is therefore

n-channel $\quad V_p < 0$

p-channel $\quad V_p > 0$

The input resistance of a junction FET is very high since it is looking into a reverse-biased diode. Further, the JFET is a voltage operated device unlike the bipolar transistor which is current operated.

The junction gate FET is one of two basic types of this form of transistor and because it requires a voltage to be applied to the gate terminal to turn it 'off', it is said to operate only in the 'depletion-mode'.

2.6.2   The insulated-gate FET

It is the high input resistance of the junction-gate FET which makes it a particularly attractive device in many applications. If an extremely high input resistance is necessary, another type of FET is available. This is the metal-oxide FET or MOSFET, sometimes also referred to as a MOST or insulated-gate FET, or IGFET. This device differs from the JFET by having its gate electrode isolated from the channel by a thin layer of insulation, usually silicon dioxide, as shown in Fig. 2.14. Here the planar construction creates two n-regions within a p-type substrate, the n-regions being heavily doped to make them rich in electrons. The whole surface of the material is then covered with an oxide layer except for the two contract areas to the n-type regions, source and drain respectively. A metallic gate electrode is then deposited on the surface of the oxide layer between the two n-regions. In this way the gate is

45

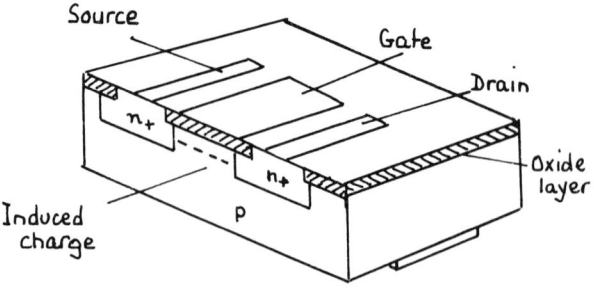

Figure 2.14

completely insulated from the substrate. The input resistance of the
device is consequently very large, typically of the order of 10^6 to
10^8 megohms.

When a positive potential is applied between gate and source, the
gate electrode and the substrate surface behave as the plates of a
capacitor separated by the oxide dielectric, hence a negative charge
is induced in the substrate surface between the n-regions. Fig.
2.15(a) shows the situation in the FET with a very small positive
gate voltage on the source and drain. The induced negative charge
creates a depletion region beneath the gate electrode and this forms

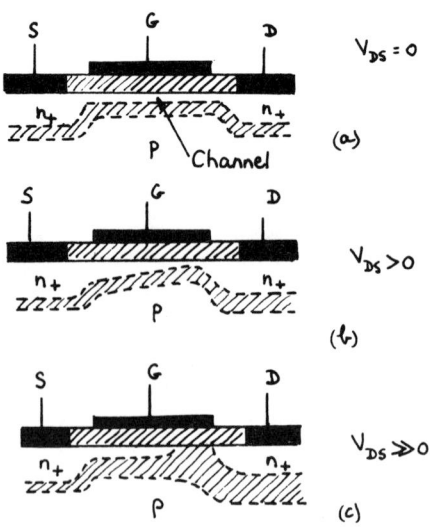

Figure 2.15

a continuous depletion layer between the drain and source regions. If the gate voltage is increased slightly the region beneath the gate becomes sufficiently negative to appear like n-material relative to the p-type substrate in which it is formed. This effect is known as an inversion. This inversion channel extends from the source to the drain, both heavily doped n-regions. As the gate voltage is increased further, the conduction channel increases in depth; hence the channel width is under the control of the gate voltage as it was in the case of the junction-gate FET.

If a positive voltage is now applied to the drain, a current will flow along the inversion channel as shown in Fig. 2.15(b) but the depletion region will be distorted by the effect of the voltage gradient which now exists between drain and source. This gradient will tend to cancel the field produced by the gate bias. As V_{DS} is increased further this cancellation effect will finally become sufficient to prohibit the formation of the inversion channel at the drain end and the channel will pinch off, see Fig. 2.15(c). Saturation current then flows at a level that is substantially independent of further increases in drain voltage. In the absence of an inversion channel there would be no drain current; thus the pinch-off voltage V_p (or threshold voltage V_T as it is sometimes known) is that voltage which when applied to the gate just produces inversion in the substrate.

With this type of MOSFET, a positive voltage must be applied to the gate to provide a conduction channel and turn the transistor on. An increased field increases or 'enhances' the channel width and consequently the magnitude of the drain current. This form of operation is known as the 'enhancement mode'. Figure 2.16 shows the out-

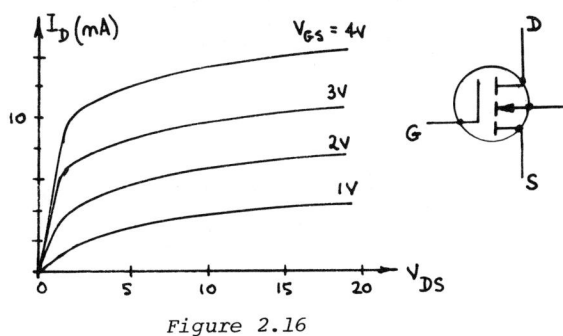

Figure 2.16

put characteristic for an enhancement n-channel MOSFET together with
its circuit symbol. Because these are normally OFF devices the symbol incorporates broken lines to represent channels that are normally
open-circuit. The substrate is marked through in most cases but normally no external connection is or can be made to it.

An alternative form of MOSFET can be constructed by introducing
a lightly doped layer linking source and drain. In the p-substrate
model we have been considering, this added layer would be n-type.
In this way, a conducting channel is present even when V_{GS} is zero,
and the application of a positive potential to the gate will enhance
the channel and give increasing drain current until pinch-off is
reached. But now, unlike the enhancement mode FET illustrated in
Fig. 2.15, the application of a negative voltage to the gate will
also have an effect on drain current; it will tend to deplete the
channel which has been deliberately diffused into the substrate surface, and by increasing V_{GS} sufficiently in the negative sense, the
transistor can be turned off, exactly as was the case for the
junction-gate. The device is consequently now operating in the
depletion mode. Figure 2.17 shows the characteristics and circuit
symbol for this enhancement or depletion mode MOSFET. This is a
normally ON device and the channel is indicated by a full line.

2.7   THE TRANSFER CHARACTERISTICS

In the field-effect transistor, the output current is controlled by
the input voltage. Within the normal operating region, that is, over
the saturation region of the output characteristics, drain current

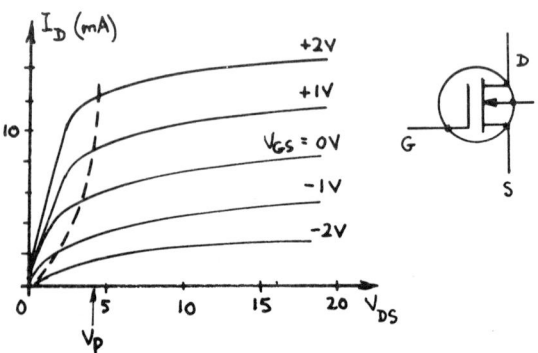

Figure 2.17

I_D is substantially independent of the drain-source voltage V_{DS}. Over the same region, however, I_D is highly dependent on the gate-source voltage and the transfer characteristics showing the relationship between I_D and V_{GS} for all three forms of the FET are shown in Fig. 2.18. These illustrate that the JFET and the depletion MOSFET are normally 'on' devices, while the enhancement MOSFET is a normally 'off' device. The value of I_D for zero gate voltage is designated I_{DSS}.

From experimental analysis it has been shown that the transfer characteristics approximate to a parabolic (or square law) form in all three cases. For the JUGFET and the depletion MOSFET, the drain current over the operating region is closely related to V_{GS} by the expression

$$I_D = I_{DSS} \left[1 - \frac{V_{GS}}{V_P} \right]^2 \tag{2.5}$$

and (for the depletion MOSFET) both positive and negative values of V_{GS} are allowed.

For the enhancement MOSFET, the corresponding relationship can be shown to be

$$I_D = k(V_{GS} - V_T)^2 \tag{2.6}$$

where V_T is the turn-on or threshold voltage and k is a constant.

Manufacturers usually specify I_{DSS}, V_p and V_T and the d.c behaviour of FETs is then readily deduced.

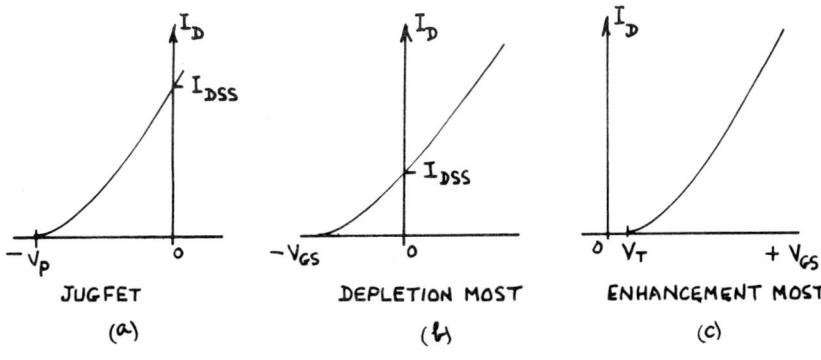

JUGFET   DEPLETION MOST   ENHANCEMENT MOST

(a)   (b)   (c)

Figure 2.18

2.8   TRANSIENT EFFECTS IN p-n JUNCTIONS

Under reverse bias conditions, the densities of both carriers in the vicinity of a p-n junction are lower than in the unbiased case and this depletion of carriers extends over a greater distance. As a result the space charge layers (from the bound charges) are stronger and thicker than they are with no bias. Consequently the electric field acting across the junction is stronger under reverse bias. The integral of this field, the potential difference across the junction, is correspondingly greater, as it must be to include the applied voltage. All this basically recapitulates our earlier notes on the biased junction.

   As the reverse voltage is varied the thickness and strength of the space charge layers will vary, thus there is charging and dis-charging of the opposite faces of the junction and the junction exhibits a capacitance. This capacitance varies with the applied voltage and forms the basis of the 'varactor' diode. Figure 2.19 shows the variation of the total charge, Q, the field strength, ε, and the dielectric thickness, ℓ, with applied voltage V in an ordinary capacitor. C is, of course, constant. Diagram (b) shows the same things for a junction. In this latter case Q, ℓ and ε vary as $\sqrt{V}$ while C varies as $1/\sqrt{V}$. Thus Q × ℓ and Q/C both vary as V as must be true. In these relations V is the total potential drop across the junction i.e. the applied voltage plus the contact potential.

   When the junction is forward biased the hole density in the n-type material and the electron density in the p-type material is greatly increased near the junction. If the bias is suddenly re-versed, these carriers will be pulled back across the junction and

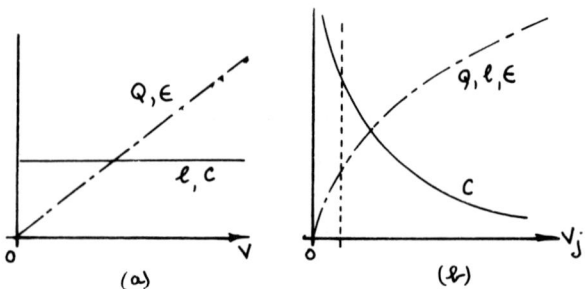

Figure 2.19

the reverse current will not decay to its steady saturation level
until this action, together with the normal recombination process,
has decreased the carrier densities to a low value. As a consequence,
under the sudden application of reverse bias after forward conduction has been taking place, a pulse of reverse current will occur
as shown in Fig. 2.20. This effect is known as 'storage' because the
high n- and p-carrier densities in the material of opposite types
represents a store of carriers which will provide reverse conduction
and which must be exhausted before the full back resistance can
develop. Thus, in addition to the junction capacitance just considered, the junction also exhibits a 'diffusion' capacitance, represented in the form of the stored charges.

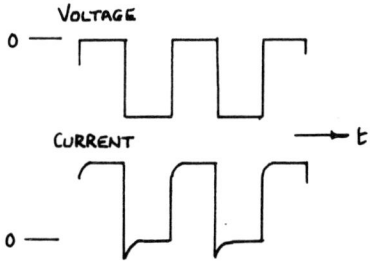

Figure 2.20

The effects of junction capacitance and carrier storage (diffusion capacitance) is to limit the high frequency performance of
junction diodes and transistors. In the bipolar transistor the collector capacitance is of importance because the collector-base
junction is reverse biased and the collector impedance is high. For
a high gain, the collector should look into a high impedance load;
the junction capacitance is however in shunt with the load and this
limits the gain-bandwidth product. (This product will be discussed
in a later chapter.)

The diffusion capacitance in a transistor results almost entirely
from the injection of the carriers from the emitter into the base.
The situation is best pictured in a diagram, see Fig. 2.21. In the
base there is a concentration gradient and the carriers tend to drift
from the region of high density to that of low density rather than
the other way about. The difference is the net current one way.

51

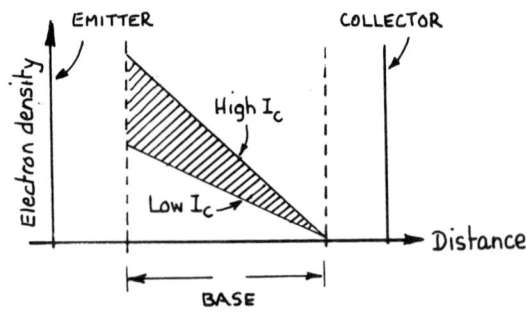

Figure 2.21

At the collector face the concentration is low; as we move towards the emitter the density increases with a gradient proportional to the collector current. The base diffusion of the carriers represents a stored charge which is neutralized by electrons (for an n-p-n device) flowing out of the base terminal. Thus to 'change' the collector current requires that the base be charged with a number of carriers proportional to the shaded area shown in Fig. 2.21. Each change of base current has, therefore, two components: one component contributes to the stored charge in the base region as if a capacitance were connected across the base-emitter junction, and the other component produces the desired change in the collector current. At high frequencies where the base variations are very rapid, most of the emitter current is 'lost' in the charging and recharging operation and the collector current changes are very small, i.e. the current gain is small.

The frequency at which the gain has fallen by 3 dB from its d.c. value is known as the 'alpha cut-off' frequency. In early transistors a few megahertz was considered good, but frequencies in excess of 600 MHz are now readily realised.

The effect of alpha cut-off (ω_α) is essentially different from that of collector capacitance in that the latter limits only the gain-bandwidth product and places no restrictions on the midband frequency, whereas alpha cut-off sets an upper limit to the frequency at which the gain may be had at a given bandwidth.

2.9   A LABORATORY EXPERIMENT

This experiment enables α_E to be determined for a series of values
of emitter current. Hence the variation of α_E with changes in emitter current can be shown.

Connect up the circuit of Fig. 2.22. Set the low-frequency
signal generator to resonate with the tuned circuit by any suitable
means — this should be at a frequency of about 2200 Hz. With switch
S_1 set to position A and the emitter current set to 1 mA, adjust
the generator output so that the a.c. millivoltmeter reads 10 mV.
Change switch S_1 over to position B, close switch S_2 and record the
voltage on V_0. Deduce for yourself that this output voltage is
numerically equal to the value of α_E, i.e. 50 mV would correspond
to $\alpha_E = 50$.

Repeat the above process for a number of emitter currents and
plot a graph of I_E against α_E. A suitable range of I_E is from
0.1 mA to 5 mA. Ensure that there is no significant temperature
change during the experiment.

A transistor with a relatively low α_E (say < 100) should be used
in this experiment.

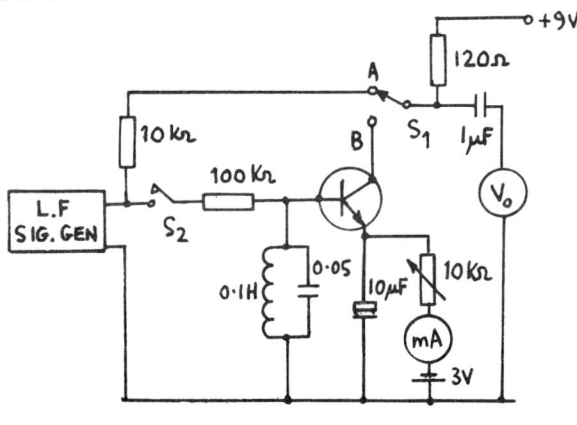

Figure 2.22

PROBLEMS 2

1. Explain the function of the emitter, base and collector regions
in a bipolar transistor.

2. What effect, or effects, does the width of the base region have

on transistor operation?

3. Why is the base region more lightly doped than either of the emitter or collector regions in a bipolar transistor?

4. Why would not two diodes connected back to back provide transistor operation?

5. Explain the terms (a) emitter efficiency, (b) base transport factor. What does the product of these terms represent?

6. Differentiate between I_{CBO} and I_{CEO}. What is the relationship between them?

7. I_{CBO}, I_{CEO} and I_{EBO} all symbolise transistor leakage currents. Sketch circuits that you might use independently to measure each of these currents in an n-p-n transistor.

8. A germanium transistor has $\alpha_B = 0.99$ and $I_{CBO} = 12$ μA. What will be the value of I_E if the base is disconnected?

9. Show that $$\alpha_B = \frac{\alpha_E}{1 + \alpha_E}$$

10. A transistor having $\alpha_B = 0.995$ and $I_{CBO} = 5$ μA has $I_B = 20$ μA. What is the collector current?

11. Prove that three currents I_E, I_C and I_B flow in the ratio $1 : \alpha_B : 1 - \alpha_B$, whatever the configuration.

12. Why does an increase in V_{CB} lead to the effect shown in Fig. 2.7(a)?

13. A transistor has its terminals marked as P, Q and R and is operating as a conventional amplifier. With no input signal, you take measurements and note that terminal P is 12 V positive with respect to terminal Q, and 0.6 V positive with respect to terminal R. Justifying your answers, assign the emitter, base and collector regions to the appropriate terminals P, Q and R. Is the region represented by P, p- or n-type semiconductor?

14. Describe briefly the fabrication of an n-p-n transistor using the planar method.

15. A transistor is connected as shown in Fig. 2.23. Assign the proper polarities to V_{BE} and V_{BC} for normal transistor operation.

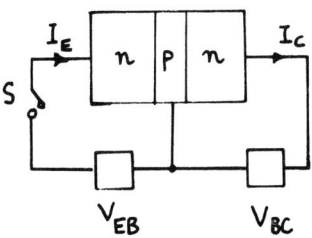

Figure 2.23

Sketch a curve of I_C against ± values of V_{BC} with switch S (a) open, (b) closed and I_E = 1.5 mA.

16. A transistor with α_E = 100 and I_{CBO} = 5 μA is connected in common-emitter mode and you measure a collector current of 1 mA. Calculate I_E, I_B, α_B and I_{CEO} under this condition.

17. What is meant by (a) diffusion capacitance, (b) carrier storage, (c) alpha cut-off frequency, as applied to a bipolar transistor?

18. Describe the operating principles of an n-channel junction-gate FET and sketch a cross-section of the device at pinch-off.

19. Explain the operation of (a) a depletion, (b) an enhancement type MOSFET. Is a JFET device a depletion or an enhancement type FET?

20. At pinch-off, why does the drain current of a FET not fall to zero?

21. Using equation (2.5) in the text, show that the gradient of the transfer characteristic of a depletion type FET is given by

$$\frac{-2I_{DSS}}{V_p} \left[1 - \frac{V_{GS}}{V_p} \right]$$

3 Fundamental principles of amplifiers

3.1   CLASSIFICATION OF AMPLIFIERS

Electronic amplifiers are divided into a great number of classifica-
tions and sub-classifications, but we shall be interested only in the
broad division of what are known as 'linear' amplifiers, designed to
handle 'analogue' signals which carry information in terms of their
waveforms and amplitudes; and switching amplifiers designed to handle
'digital' signals where the information is carried in the pattern of
pulse trains. Only these types find general use in integrated
circuit form.

   Linear amplifiers are sub-divided into small-signal and large-
signal forms. The small-signal category contains those amplifying
stages designed to deal with the output signals from such trans-
ducers as pick-up heads, tape heads and microphones. Such signals
have amplitudes lying within the range of a few millivolts to a few
volts. The large-signal category covers the range of what are also
known as 'power' amplifiers. These amplifiers are designed to provide the power necessary to operate loudspeakers, recorders, motors
and other such devices. Special problems arise in the design of
these amplifiers that are not found in the small-signal category.
For one thing, high conversion efficiency is necessary or significant
power is wasted in the system, and for another, the high signal
levels involved result, unless proper care is taken, in considerable
distortion. Discrete transistor circuitry is also much more significant in power amplification.

   The division between small-signal and large-signal amplifiers is
not clear cut; an amplifier taking its input from a recorder head (a

pre-amplifier as it is usually called) is undoubtedly a small-signal
amplifier, and an amplifier driving the loudspeaker system at the
local disco is undoubtedly a large-signal amplifier. However, there
may well be an amplifier stage (or stages) connected between the pre-
amplifier and the power amplifier. This would be accepting relatively small signals from the pre-amplifier but providing an output
sufficiently large to drive the power amplifier. In the context of
classification it might be difficult to fit this amplifier precisely
into either category.

Amplifiers may also be categorised as tuned or untuned. Tuned
amplifiers employ selective resonant circuits and operate only over a
relatively narrow frequency band, usually located above some 50 kHz.
They are also known as narrow-band or selective amplifiers and typical
examples are the aerial tuning systems of radio receivers or the
intermediate-frequency amplifiers found in superheterodynes. Untuned
amplifiers are generally broad-band and examples are audio amplifiers
having a frequency range of typically 20 Hz to 20 kHz, and video-
amplifiers found in television receivers which cover a range from a
very few hertz (d.c. in many cases) to an upper limit of 5 MHz or
more.

3.1.2   Operating classifications

Amplifier stages are divided into three main categories according to
the location of the operating point on the dynamic transfer characteristic of the amplifying device concerned.

In a Class-A amplifier, the input signal causes output current to
flow throughout the whole 360° of the input cycle. Nearly all untuned small-signal amplifiers (and some power amplifiers) are
operated under Class-A.

A Class-B amplifier has the operating point positioned so that
the output current is very small in the absence of an input signal
and flows only for about 180° of the input cycle when a signal is
applied. Most untuned power amplifiers operate under Class-B conditions.

A Class-C amplifier has the operating point positioned so that
output current flows for only part of one-half cycle of the input
signal (usually about 120°). These amplifiers are invariably tuned

amplifiers operating at high frequencies.

3.2   AMPLIFICATION AND DISTORTION

Amplifier gain is a measure of the voltage, current or power amplification provided by the amplifier. We define

$$\text{Voltage gain } A_V\underline{/\theta} = V_o/V_i = A(\cos\theta + j\sin\theta)$$
$$= A\varepsilon^{j\theta}$$

so that A_V is the complex ratio of two phasors and A and θ are func-
tionally related. If A and θ were constants, independent that is of
the signal amplitude and frequency, the output signal would be a
replica of the input signal and the amplifier would be linear. This
would be an ideal situation but real amplifiers are not so co-
operative. Invariably there will be some degree of re-shaping and
the output will, in some way or other, differ from the input waveform; in other words, distortion will occur.

3.2.1   Amplitude distortion

Amplitude distortion arises if A is not constant. This kind of distortion comes about when the input to the amplifier is excessive and
the signal traverses sections of the amplifier characteristic which
are non-linear. The output then contains frequency components (harmonics) of varying phases and amplitudes that were not present in the
input. Figure 3.1 illustrates the way, for example, that a second
harmonic component is introduced, assuming that the characteristic
follows a square law. Amplitude distortion can be reduced by operating with small signal inputs. However, there is a limit to the
smallness of the signal because of the inherent noise introduced
by the amplifier; once the noise becomes comparable with the signal,
the output is effectively useless. Noise has the greatest implica-
tions in input stages where signal levels are naturally very small.

3.2.2   Frequency distortion

Frequency distortion arises if the gain A is a function of frequency
so that all frequencies within the dynamic range of the amplifier
are not amplified equally. An amplifier suffering from frequency
distortion might have a response curve of the shape shown in Fig. 3.2.

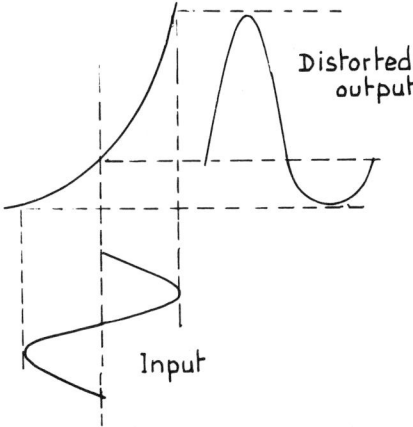

Figure 3.1

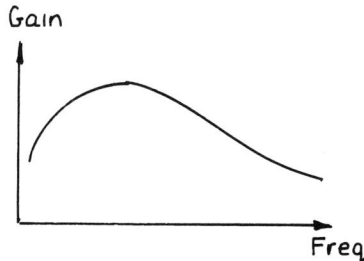

Figure 3.2

Here there is a loss of gain at high frequencies and a considerable increase in gain at low frequencies. The output signal clearly
will be distorted in waveform because the many harmonic components
of different frequencies of which it is made up will not be amplified
in their proper relative proportions. The ear is very sensitive to
frequency (and amplitude) distortion. No amplifier is completely
free from frequency distortion.

3.2.3   Phase distortion

If θ is a function of frequency, the relative amplitudes of the signal harmonic components may not be affected but the relative phase
positions are altered. As Fig. 3.3 shows, when the input signal contains a harmonic and that harmonic is shifted in phase relative

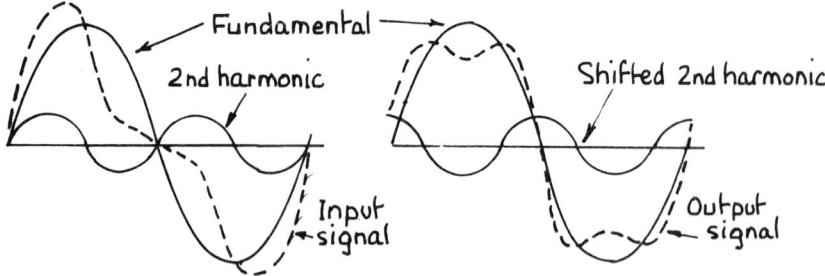

Figure 3.3

to the fundamental, the output wave can be totally different in shape to the input wave. Such phase distortion is not particularly significant in audio amplifiers as the ear is not sensitive to it. Both phase and frequency distortion are caused by frequency dependent circuit elements (inductors and capacitors) and by the variations in gain and phase shift exhibited by active devices such as transistors, particularly at very low and very high frequencies.

3.3   BASIC PRINCIPLES OF AMPLIFICATION

The essentials of an amplifier are that a small signal input is made to control power which is delivered by an energy source so that it may be dissipated in a load. Of all the possibilities, controlled variable dissipation amplifiers are the most important and find con- siderable use in integrated circuit form. The load is connected in series with the active device and the energy source, see Fig. 3.4. If the voltage drop across the device is V and the current flowing is I, then the power dissipated in the series load resistance R_L is

$$P_L = I^2 R_L = I(E - V)$$

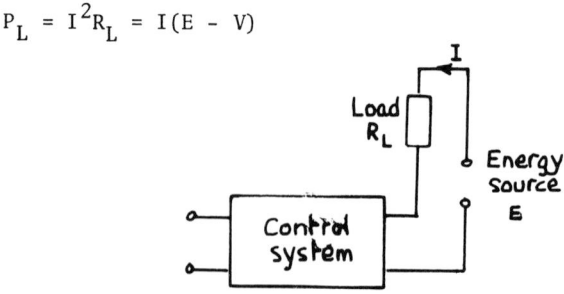

Figure 3.4

where E is the power source e.m.f. This power is controlled by the
input signal. The dissipation device itself wastes part of the power
drawn from the source which is

$$P_D = IV = I(E - IR_L) = EI - P_L$$

The power efficiency is therefore

$$\eta = \frac{P_L}{P_L - P_D}$$

and is necessarily less than 1. The wasted power is dissipated as
heat and causes a temperature rise in the active device.

Transistors, both FET and bipolar, and thermionic valves are controlled variable dissipation devices. Thermionic valves are now
generally obsolete although they still retain superiority in high
power fields, and a great deal of industrial equipment incorporating
valves can still be encountered. The FET has characteristics most
similar to thermionic valves but its smaller bulk, greater reliability
and lower operating voltages give the FET unchallenged superiority
over the valve.

A fundamental current amplifier in common-emitter mode is shown
in Fig. 3.5(a) together with its collector characteristics. The
transistor which is the controlled dissipation device is connected
in series with a load resistor R_L and an energy source, a battery
with terminal voltage V_{cc}. A steady (or quiescent) base current I_B

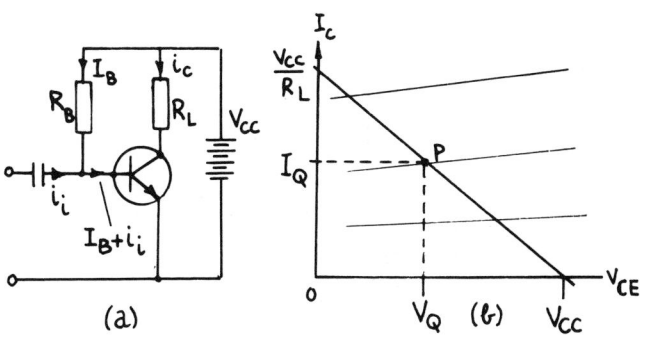

(a)   (b)

Figure 3.5

is established by way of resistor R_B and the base-emitter junction is
thereby forward biased. The input signal which we assume is sinusoidal, is applied in parallel with the emitter biasing circuit. The
reverse bias on the collector is maintained by V_{CC} and most of this
voltage appears across the collector-base junction since the voltage
drop across the forward biased emitter junction is small, typically
0.6 V for a silicon transistor. A signal current i_i will cause a
variation in base current i_B, where $i_B = I_B + i_i$, and this will produce a variation in collector current i_C. The varying component of
i_C will constitute the amplified output current

3.3.1   The load line

The static or d.c. load line is a line drawn across the collector
characteristics of a transistor and represents the locus of all
possible operating points for the transistor and its associated circuitry. A load line is seen on the characteristics of Fig. 3.5(b).
Clearly the position of this line relative to the co-ordinate axes
depends upon V_{CC} and R_L. From the circuit diagram

$$V_{CC} = v_{CE} + i_C R_L$$

so that

$$i_C = -\frac{1}{R_L} v_{CE} + \frac{V_{CC}}{R_L} \tag{3.1}$$

where i_C and v_{CE} are the dependent and independent variables respectively.

This is seen to be the equation of a straight line in gradient
and intercept form; the gradient is $-1/R_L$ and the vertical intercept
is V_{CC}/R_L. The operating point must lie on this line as well as on
the characteristic curve for $i_B = I_B$. The intersection of the line
and each of the characteristic curves represent graphical solutions
of equation (3.1) and define the 'instantaneous' values of i_C and
v_{CE} for any given value of the input signal i_i. At the operating
point, shown as P, the steady or quiescent of collector
current and voltage are respectively denoted by I_Q and V_Q. The
signal output current i_o is the a.c. component of i_C, the total

current in R_L.

The actual selection of an operating point is subject to a number of factors, amongst which are the maximum signal excursions likely to be applied to the amplifier, the available supply voltage and what degree of distortion (if any) can be tolerated. For linear Class-A amplification it is general to make the quiescent collector voltage about one-half of V_{CC}; this gives an equal chance for an output voltage swing in either direction, but there is nothing at all to prevent other points being selected and used in a particular circumstance.

Graphical analysis is usually reserved for large-signal stages in any event, but we are using the method here before considering large-signal amplifiers as a means of analysing the effects of leakage current and variations in α gain figures on the stability of the operating point.

The load line is located at its lower end at a point on the V_{CE} axis equal to V_{CC}, obtained by setting $i_C = 0$ in equation (3.1). Its upper end intersects the current axis at V_{CC}/R_L, obtained by setting $V_{CE} = 0$. Only these extreme points are required to establish the load line.

3.4   JUNCTION TRANSISTOR BIASING

The operating point can shift from a desired location because of two major causes, and we recapitulate:

(a) the collector-base leakage current I_{CBO} which is temperature dependent,

(b) the common-emitter current gain figure α_E which is subject to wide variation from unit to unit.

As an example of the problem, consider Fig. 3.6. The operating point is located at P. The effect of a time-varying base current of about 100 μA peak value can now be analysed in terms of the load line and the collector characteristics of the transistor. For with the operating point at P (the no-signal or quiescent condition) the excursions in base current i_B permissible about this point which will result in proportional changes in collector current are approximately those between which the load line intersects the characteristics in equal spacing on either side of P. At no time should the transistor

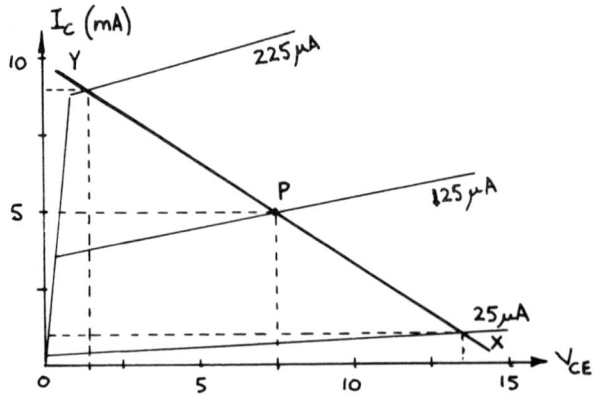

Figure 3.6

be swung beyond collector current cut-off at the end X of the line
nor should the base input be sufficiently large to work into the
curved portions of the characteristics at the Y end of the line. It
is seen from the diagram that the base current variation of 100 µA
peak is causing a collector current peak variation of about 4 mA
superimposed on the quescent value of 5 mA. The corresponding excursion of collector voltage is then about 6 V peak about the quiescent
value of 7.5 V. This is standard Class-A operation.

Suppose now that a transistor of the same type but having a different current gain figure is substituted into the circuit; or there
has been a relatively large change in the ambient temperature. The
operating condition may now be as shown in Fig. 3.7. Although the

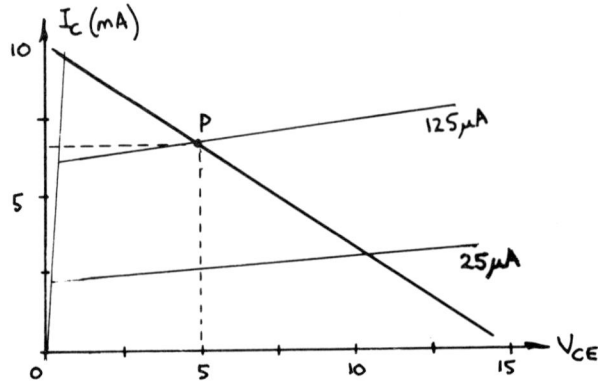

Figure 3.7

operating point is still at I_B = 125 μA, the position of P relative
to the quiescent collector current has changed; the characteristics,
in other words, have moved relative to the co-ordinate axes. The
introduction of the same base signal current now results in a distorted output, the stage being driven into saturation. Figure 3.7
differs from Fig. 3.6 because it represents either a higher gain
transistor or an elevated temperature. The objective in the biasing
of discrete transistors is therefore to establish a proper operating
point and to MAINTAIN it despite variations in temperature and
variations among individual devices of a similar type. We have already discussed thermal instability and the effects of device tolerance in Section 2.5 of Chapter 2.

3.4.1   Stability factor

As the shift in the operating point is relative to the co-ordinate
axes and not the base current lines, the maintaining of a constant
base bias current does not provide a solution to the problem. If,
however, the operating point can be maintained at some fixed value
of I_C, i.e. if some point along the co-ordinate axes is held constant,
the problem is eased. It is general, then, to evaluate the method
of biasing in terms of the stability of the collector current.

A stability factor S has been defined and is expressed as

$$S = \delta I_C / \delta I_{CBO}$$

This tells us the ratio of the 'change' in collector current δI_C
caused by the change δI_{CBO} in leakage current. S must clearly be
greater than or equal to unity and the closer to unity the better for
circuit stability.

Refer now to the common-emitter circuit shown in Fig. 3.8. A
separate supply V_{BB} is shown for biasing and an emitter resistor R_E
is included in the emitter lead. The voltage drop across R_E tends
to reverse-bias the emitter junction but the positive potential at
the base provided by V_{BB} by way of R_B sets the base voltage so that
an overall forward bias is applied. R_L is chosen so that the voltage
drop across it due to the quiescent collector current leaves the collector at a specified voltage with respect to the emitter (not the
earth line), usually about one-half of V_{CC}. Applying Kirchhoff to

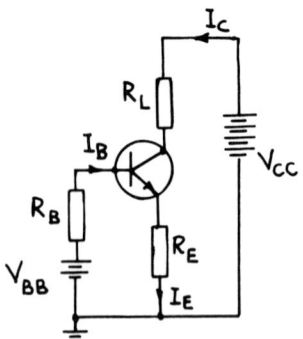

Figure 3.8

the output loop we obtain

$$I_C R_L + V_{CE} + R_E(I_C + I_B) - V_{CC} = 0$$

Since $I_C \gg I_B$ this can be simplified to

$$V_{CE} + I_C(R_L + R_E) = V_{CC} \qquad (3.2)$$

Knowing V_{CC}, V_{ce}, I_C $(= I_Q)$ and R_E enables R_L to be evaluated.
   Applying Kirchhoff around the base loop yields

$$R_B I_B + R_E(I_C + I_B) = V_{BB} - V_{BE} \qquad (3.3)$$

Now

$$I_C = \alpha_E I_B + I_{CEO}$$

Therefore

$$I_B = \frac{I_C - I_{CEO}}{\alpha_E}$$

Substituting this value of I_B into equation (3.3) we have

$$R_B \left[\frac{I_C - I_{CEO}}{\alpha_E} \right] + R_E \left[I_C + \frac{I_C - I_{CEO}}{\alpha_E} \right] = V_{BB} - V_{BE}$$

and rearranging

$$\frac{I_C}{\alpha_E} [R_B + R_E (1 - \alpha_E)] - \frac{I_{CEO}}{\alpha_E} [R_B + R_E] = V_{BB} - V_{BE}$$

Differentiating throughout

$$\delta I_C [R_B + R_E (1 + \alpha_E)] - \delta I_{CEO} [R_B + R_E] = 0$$

Therefore

$$\frac{\delta I_C}{\delta I_{CEO}} = \frac{R_B + R_E}{R_B + R_E (1 + \alpha_E)} \qquad (3.4)$$

Since

$$I_{CEO} = (1 + \alpha_E) I_{CBO}$$

$$\frac{\delta I_C}{\delta I_{CBO}} = S = \frac{(R_B + R_E)(1 + \alpha_E)}{R_B + R_E (1 + \alpha_E)} \qquad (3.5)$$

This gives us an expression for the stability factor in terms of R_B, R_E and α_E. By the substitution $\alpha_E = \alpha_B / (1 - \alpha_B)$, this can be written in the alternative form

$$S = \frac{R_B + R_E}{R_E + R_B (1 - \alpha_B)} \qquad (3.6)$$

By consideration of the two extreme cases $R_E = 0$ and $R_B = 0$ in either of these relationships, the conditions for stability can be assessed.
1. For $R_E = 0$, equation (3.5) reduces to $(\alpha_E + 1)$. This makes S very much greater than unity and is the worst possible condition for stability. We have, in fact, pure 'common-emitter conditions', under fixed bias.
2. For $R_B = 0$, equation (3.6) reduces to unity. This represents perfect stabilization, the 'pure common-base condition'. This ideal condition cannot be implemented in practice since a d.c. short-circuit would be placed across the input terminals.

Clearly, all intermediate values of R_E and R_B give values of S which lie between unity and $(\alpha_E + 1)$. It can be shown that when S is large compared with unity but small compared with $(\alpha_E + 1)$, S is given approximately by the ratio R_B/R_E. Hence the stability factor will be reduced by increasing the emitter resistance R_E and reducing the base resistance R_B. It is seen from equation (3.6) that to minimise the effect of the $(1 + \alpha_E)$ term we must arrange for R_E to be very much greater than $R_B(1 - \alpha_B)$, or R_B/R_E very much less than α_E. This is feasible in some degree if the bias supply V_{BB} is a LOW voltage source separate from V_{CC} as Fig. 3.8 indicates. In practice V_{CC} is used for both supplies in the interests of simplification, see Fig. 3.9, and the inequality cannot then be satisfied since R_B must, of necessity, be large and its use will tend to set a 'constant' base current, something we are trying to avoid.

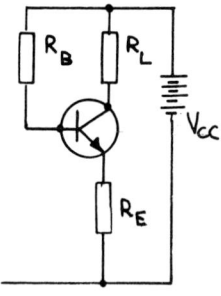

Figure 3.9

The circuit system of Fig. 3.9 is sometimes used in practice because it entails a minimum of components and the input resistance is not particularly affected by the presence of R_B which, although being in shunt with the input terminals from the signal viewpoint, has a sufficiently large value to make such shunting negligible. The circuit also enables a single battery to be used. Basically the method is satisfactory only where the ambient temperature is constant or nearly so, and the transistor is operated so well within its ratings that any internal heating is insignificant.

Example 3.1   A silicon transistor is used in the circuit of Fig. 3.10 with V_{CC} = 9V, R_L = 3.9kΩ and I_C set to an operating level of 1 mA. Taking the least value of α_E to be 50, assess suitable values for R_E, R_B and V_{BB} for a stability factor of 5.

The voltage drop across R_L is 3.9 V. Allowing the same for V_{CE}

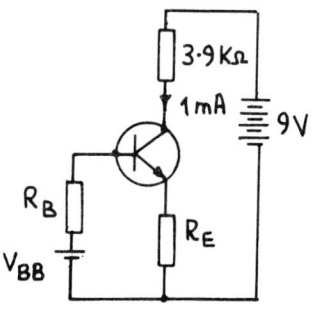

Figure 3.10

so that the operating point will be roughly at the centre of the load line, the drop across R_E will be 1.2 V. Hence R_E will be 1.2 kΩ since $I_E \simeq I_C$. From equation (3.5) we then obtain

$$5 = \frac{(R_B + 1.2)(50 + 1)}{R_B + (1.2 \times 51)}$$

and so

$$R_B = 5.32 \text{ k}\Omega$$

The voltage drop across R_E is 1.2 V and taking V_{BE} = 0.6 V, the base-emitter voltage added to that across R_E gives a total of 1.8 V between base and earth. But $I_B \simeq I_C/\alpha_E$ = 1/50 mA, hence the voltage drop across R_B is approximately 5.32/50 = 0.1V. Therefore

$$V_{BB} \simeq 1.8 + 0.1 = 1.9 \text{ V}$$

Preferred values for R_E and R_B in a practical circuit would be 1.2 kΩ and 5.6 kΩ. Make a note that we have worked in mA and kΩ, so avoiding the use of self-cancelling 10^3 and 10^{-3} terms.

This problem illustrates the fact that although a supply of 9 V is AVAILABLE, only 1.9 V is REQUIRED for the bias supply; it is not possible to return R_B to the 9 V line without greatly increasing its value above 5.6 kΩ and this would invalidate the agreed value for the stability factor.

3.4.2   Other bias arrangements

The use of a separate bias supply is obviously inconvenient and some means must be found of obtaining the base bias from the V_{CC} rail without at the same time introducing a poor stability factor.

Some improvement can be obtained by connecting R_B between collector and base as shown in Fig. 3.11. Because the collector voltage is lower than V_{CC} the value of R_B is necessarily smaller than it would be if it was returned directly to the V_{CC} line. Any temperature-produced change in the static characteristics results in a change in the no-signal collector potential and the base current changes correspondingly because V_{CE} is in effect supplying I_B. This, in turn, modifies the collector current in such a direction that the original change is nullified. Thus with this arrangement a constant I_B is not being set because I_B is dependent upon the d.c. output variations.

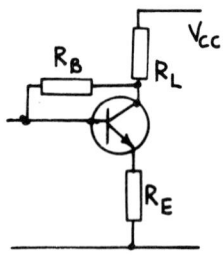

Figure 3.11

A further, and very commonly used modification, is illustrated in Fig. 3.12(a). Here an additional resistor R_1 is connected between base and earth, the former R_B now designated R_2 being returned to the V_{CC} rail. Forward bias for the emitter junction is now obtained by way of a potential divider system. As before, the voltage drop across R_E tends to reverse-bias the junction but the voltage divider sets the base voltage sufficiently positive to ensure overall forward conduction. Any tendency for I_C to increase due either to an increase in I_{CBO} or α_B is linked to a corresponding increase in I_E. As a result the p.d. across R_E tends to increase which in turn reduces V_{BE}, causes the base bias to fall, and opposes the original change in I_C. The additional resistor R_1 can be selected so that the 'equivalent' base resistance (R_B in Fig. 3.8) is considerably reduced; an

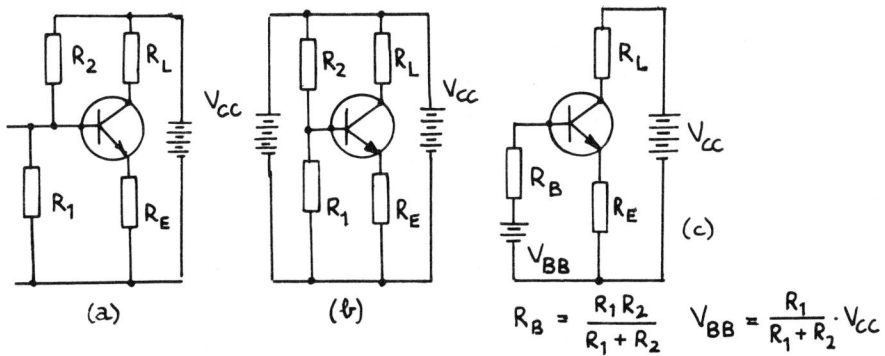

Figure 3.12

improvement in the stability factor is therefore possible. The problem is to choose R_1 and R_2 so that the correct equivalent V_{BB} is
obtained in conjunction with the required equivalent R_B.

To analyse this, the circuit can be reduce to its Thevenin form
by way of Fig. 3.12(b). The resistance seen looking into terminals
A and B by the transistor is R_1 and R_2 in parallel, and the open-
circuit e.m.f. across these terminals is that fraction of V_{CC} given
by

$$\frac{R_1}{R_1 + R_2} \cdot V_{CC}$$

Hence the circuit reduces to the forms already discussed under Fig.
3.8 where

$$\frac{R_1 R_2}{R_1 + R_2} \quad \text{replaces } R_B \quad \text{and} \quad \frac{R_1}{R_1 + R_2} \cdot V_{CC} \text{ replaces } V_{BB}$$

See Fig. 3.12(c).

Returning now to equation (3.3), by rearrangement

$$V_{BB} - V_{BE} - I_B R_B - (I_C + I_B) R_E = 0$$

Also we may write

$$I_B = \frac{I_C}{\alpha_E} - I_{CBO}$$

71

Substituting and solving for I_C we get

$$I_C = \frac{V_{BB} - V_{BE} + I_{CBO}(R_B + R_E)}{R_E + \dfrac{R_B + R_E}{\alpha_E}}$$

(3.7)

This provides us with an equation for I_C in a self-biasing circuit where V_{BE}, α_E and I_{CBO} are those parameters which are sensitive to temperature change and manufacturing tolerance. The variation in α_E can be very large even in the tolerance bands that manufacturers provide, and additionally it increases approximately linearly with temperature. I_{CBO} as we have already noted will approximately double for every $10^\circ C$ increase above room temperature; and typically V_{BE} decreases about 2.5 mV per $^\circ C$ rise.

A study of equation (3.7) shows that the circuit will be stable if each term containing one of the above variables is made small relative to the appropriate circuit constant, that is, the resistors and the supply voltage. Approximations are quite legitimate in assessments; for $R_B > 10R_E$ we may assume that $R_B \simeq (R_B + R_E)$. Then to make I_C independent of α_E we must make

$$\frac{R_B + R_E}{\alpha_E} \simeq \frac{R_B}{\alpha_E} \ll R_E$$

(3.8)

This confirms the statement made earlier on page 68.

To make I_C independent of I_{CBO} we must make

$$I_{CBO}(R_B + R_E) \simeq I_{CBO}R_B \ll V_{BB} - V_{BE}$$

And to make I_C independent of V_{CE} we must make $V_{BE} \ll V_{BB}$.

Example 3.2   According to a catalogue, α_E for a batch of 2N2904 transistors varies from 40 to 120 and I_{CBO} may vary from 0.1 μA to 0.5 μA. If the variation in V_{BE} is from 0.5 V to 0.8 V, assess the variation in I_C for the circuit of Fig. 3.12(a) given in $R_1 = 10$ kΩ, $R_2 = 68$ kΩ, $R_E = 1$ kΩ, $V_{CC} = 15$ V.

Thevenin equivalents for R_B and V_{BB} are

$$R_B = \frac{R_1 R_2}{R_1 + R_2} = \frac{680}{78} = 8.7 \text{ k}\Omega$$

$$V_{BB} = \frac{R_1}{R_1 + R_2} \cdot V_{CC} = \frac{10}{78} \times 15 = 1.92 \text{ V}$$

The worst case occurs when $\alpha_E = 40$, $V_{BE} = 0.8$ V and $I_{CBO} = 0.1$ μA. Substituting these values into equation (3.7) above we have

$$I_C = \frac{1.92 - 0.8 + 0.1 \times 10^{-6}(9.7 \times 10^3)}{10^3 + \frac{9.7 \times 10^3}{40}}$$

$$= \frac{1.2 + 0.00097}{1242.5} \text{ A} = 0.9 \text{ mA}$$

For $\alpha_E = 120$, $V_{BE} = 0.5$ V and $I_{CBO} = 0.5$ μA

$$I_C = \frac{1.92 - 0.5 + 0.5 \times 10^{-6}(9.7 \times 10^3)}{10^3 + \frac{9.7 \times 10^3}{120}}$$

$$= \frac{1.42 + 0.00485}{1081} \text{ A} = 1.32 \text{ mA}$$

For these extreme variations, I_C shifts $1.32 - 0.9 = 0.42$ mA, a relatively insignificant amount.

Example 3.3   Using the results obtained for Example 3.1 earlier, obtain the values of R_1 and R_2 required for the modified potential divider bias circuit of Fig. 3.12(a).

From the relationships

$$R_B = \frac{R_1 R_2}{R_1 + R_2} \qquad\qquad V_{BB} = \frac{R_1}{R_1 + R_2} \cdot V_{CC}$$

we get by substitution and transformation

$$R_1 = R_B \frac{V_{CC}}{V_{CC} - V_{BB}} = 5.32 \frac{9}{9 - 1.9}$$

$$= 6.74 \text{ k}\Omega$$

$$R_2 = R_1 \left[\frac{V_{CC}}{V_{BB}} - 1\right] = 6.74 \left[\frac{9}{1.9} - 1\right]$$

$$= 25.2 \text{ k}\Omega$$

Preferred values of 6.8 kΩ and 27 kΩ could be used here.

It is of interest to pursue this example a little further. The total 'bleed' current through the divider chain is 9/31.9 = 0.282 mA which is about one-quarter of I_C. Also, the voltage at the base point is given by $I_E R_E + V_{BE}$ = 1.2 + 0.6 = 1.8 V as we have seen previously. Hence R_2 draws a current of

$$\frac{9 - 1.8}{25.2} = 0.286 \text{ mA or } 286 \text{ μA}$$

The difference, 19 μA, is the base current I_B; this compares well with the I_C/α_E value of 1/50 mA or 20 μA.

PROBLEMS

3.1 Now use equation (3.6) and the Thevenin equivalents for R_B and V_{BB} to show that the stability factor for the bias arrangement of Fig. 3.12(a) can be expressed as

$$S = \frac{R_1 R_2 + R_e(R_1 + R_2)}{R_E(R_1 + R_2) + R_1 R_2 (1 - \alpha_B)}$$

In example 3.3 above, the total divider 'bleed' current of 282 μA was about 14 times the base current of 20 μA. In general, a figure of about 10 times is aimed for. The point to bear in mind is that the bleed current is wasted current which means wasted power. In theory R_E should be large enough and R_1 and R_2 small enough to give a base potential of effectively zero d.c. resistance i.e. a 'constant voltage' source. However, R_E cannot be too large or a prohibitive proportion of the available V_{CC} supply will be lost across it; and the overall potential-divider resistance must not be so low as to load the input signal source or draw a wasteful current from the supply. It is general practice to make R_E of such a value that, for small-signal amplifiers, a drop from 0.5 to 1 V occurs across it, and to proportion the divider so that R_2 is of the order of five to ten times the value of R_E. But keep in mind that these points are generalities and are not necessarily followed in all circuits.

There is a large degree of freedom in fact in the divider form of biasing because a large number of combinations of R_1 and R_2 are possible, each combination resulting in the same operating point.

For example, taking I_C = 1 mA as the quiescent value as we did for example 3.3 above, possible pairs of R_1 and R_2 are, relative to those calculated

R_1	6.8	10	22	47	kΩ
R_2	27	39	82	180	kΩ

Each pair provides (within the preferred values selected) the same ratio, but the bleed current is obviously affected in each case. The stability factor is also changed; it would be necessary to study the circuit in more detail to find out if the particular value of S that turned up could be tolerated in the light of anticipated parameter variations.

Example 3.4   In the circuit of Fig. 3.13, the germanium transistor used has α_E = 70 which may be assumed constant. At 20°C the operating point is located at I_C = 1.5 mA, V_{CE} = 6 V. I_{CBO} has a value of 2 μA at 20°C and 20 μA at 50°C. By calculation, find the shift in the operating point brought about by a temperature rise from 20°C to 50°C, and illustrate the answer in the I_C/V_{CE} co-ordinate axes.

At 20°C I_C = 1.5 mA and V_{CE} = V_Q = 6 V. Then from the output loop equation $V_{CE} + I_C(R_E + R_L) = V_{CC}$ we get

$$6 + 1.5(1 + R_L) = 10$$

from which

$$R_L = 1.67 \text{ k}\Omega$$

The change in I_{CBO} = δI_{CBO} = 20 - 2 = 18 μA. Also the effective value of

$$R_B = \frac{10 \times 47}{10 + 47} = 8.25 \text{ k}\Omega$$

Then

$$S = \frac{(R_E + R_B)(1 + \alpha_E)}{R_B + R_E(1 + \alpha_E)}$$

$$= \frac{(1 + 8.25)(71)}{8.25 + 71} = 8.3$$

75

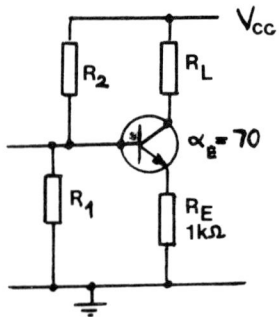

Figure 3.13

So $\delta I_C = S.\delta I_{CBO} = 8.3 \times 18 = 150$ μA or 0.15 mA. The new collector current at 50°C is therefore 1.65 mA. The new value of V_{CE} can now be determined from the loop equation used above:

$$V_{CE} = 1.65(1.67 + 1) = 10$$

Therefore

$$V_{CE} = 5.6 \text{ V}.$$

Figure 3.14 illustrates the shift in the operating point along the load line.

In practical circuit, R_E is usually bypassed with a large value capacitor. This has not been shown on any of the circuits discussed so far because its presence in no way affects the d.c. conditions. Its effect on a.c. (signal) conditions is the subject of later discussion.

We might usefully summarise at this point the biasing methods for bipolar devices:

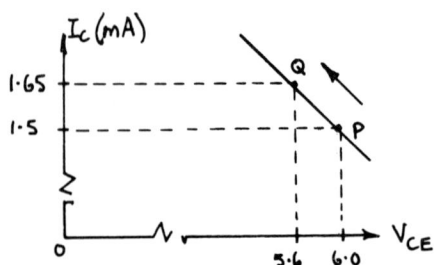

Figure 3.14

1. Fixed bias using two supplies as discussed under Fig. 3.8. This circuit arrangement has the disadvantage of two separate supplies. A fairly good stability factor can be obtained by having R_B and V_{BB} small, but R_B shunts the input and reduces the input resistance. An improvement is obtained by the inclusion of an emitter resistor.

2. Fixed bias with a single supply line as discussed under Fig. 3.9. This circuit requires a large value resistor for R_B and so the base current tends to be held constant. The shunting effect on the input is small but the stability factor is poor and the system is satis-factory only where the ambient temperature is stable and the transis-tor is operated well within its power ratings. It has the advantage of a minimum number of components and a single supply line. An improvement is obtained by returning R_B to the collector instead of the V_{CC} rail.

3. Potential-divider bias with emitter resistor as discussed under Fig. 3.12. Here the operating point stability is considerably better than that of fixed bias, only one supply line is needed and the only additional component is one resistor. Its disadvantage is the power loss due to the bleed current through the divider chain. This can generally be restricted to 1 or 2 mA for small-signal stages.

3.5   FET BIASING

Until fairly recently, field-effect transistors were used only for small-signal applications but they are now available as power ampli-fier devices. In general, the biasing requirements of FET stages are not critical and for small-signal common-source stages are com-paratively easy to determine.

3.5.1   JUGFET biasing

Since the gate potential of a depletion mode junction-gate FET must be zero or negative with respect to the source, input signals must be superimposed on a mean negative value of V_{GS}. The output character-istics of a typical JUGFET is shown in Fig. 3.15(a) with a super-imposed load line and operating point P.

To avoid the use of a second battery, a method of gate biasing is borrowed from thermionic valve techniques. A source resistor R_S is connected between the source and the earth line so that a drain current I_D flowing through it raises the potential of the source

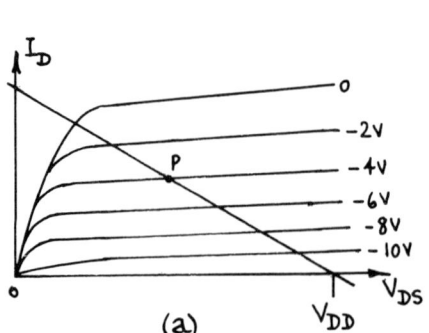
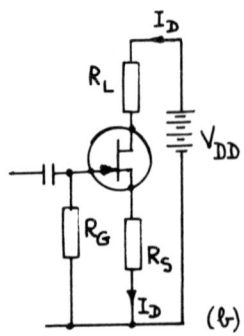

(a)

(b)

Figure 3.15

+$I_D R_S$ volts above earth. If then the gate is returned to earth
through a gate resistor R_G large enough to avoid serious shunting of
the inherently high input resistance of the FET, the gate will be
effectively at a negative potential -$I_D R_S$ volts, since the current
flow through R_G will be negligible; Fig. 3.15(b). By a proper
selection of R_S the desired biasing voltage can be readily obtained.
The method, though very simple, has the advantage that the bias is
a function of the drain current I_D. If the drain current rises the
voltage drop across R_S increases and the gate is biased towards cut-
off, thus tending to reduce I_D. This self stabilizing action makes
the amplifier very tolerant in its biasing requirements.

An improved biasing arrangement is shown in Fig. 3.16(a) where,

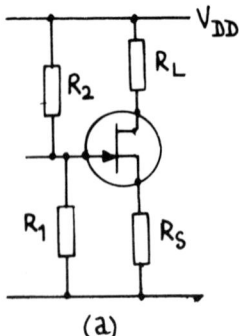
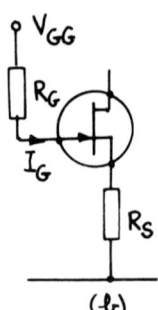

(a)

(b)

Figure 3.16

analogous to the bipolar amplifier, a potential divider made up of resistors R_1 and R_2 is adjusted to place the gate at the desired potential with respect to earth; the gate potential with respect to the source must of course still be negative. Replacing V_{DD}, R_1 and R_2 by their Thevenin equivalents, see Fig. 3.16(b) we have

$$R_G = \frac{R_1 R_2}{R_1 + R_2} \ ; \quad V_{GG} = \frac{R_1}{R_1 + R_2} V_{DD}$$

Applying Kirchhoff to the input (gate) loop gives us

$$V_{GG} - I_G R_G - V_{GS} - I_D R_S = 0$$

and since I_G will be negligible this reduces to

$$V_{GG} - V_{GS} - I_D R_S = 0$$

For a required operating point (I_D, V_{GS}) having given V_{GG} and R_L, the values of R_1, R_2 and R_S can be readily calculated as

$$R_1 = \frac{R_G R_2}{R_2 - R_G} \ ; \quad R_2 = R_G \frac{V_{DD}}{V_{GG}} \ ; \quad R_S = \frac{V_{GG} - V_{GS}}{I_D}$$

Notice that $R_S \simeq V_{GG}/I_D$ if V_{GG} is made large compared to $|V_{GS}|$; in this way the effect of any change in V_{GS} is reduced.

Example 3.5   The JUGFET illustrated in Fig. 3.16(a) is to be biased to an operating point where $I_D = 3$ mA, $V_{DS} = 6$ V and $V_{GS} = -1.5$ V. Given that $V_{DD} = 25$ V, assess the values of a biasing circuit that will be stable for appreciable shifts in V_{GS} and maintain a relatively large input resistance.

For stability we require V_{GG} to be large compared with V_{GS}, so we assume that $V_{GG} = 10$ V. To maintain a relatively large input resistance, we also assume that $R_G = 5$ MΩ. Using the above relationships yields

$$R_S = \frac{V_{GG} - V_{GS}}{I_D} = \frac{10 - (-1.5)}{3 \times 10^3} = 3.833 \text{ k}\Omega$$

$$R_2 = R_G \frac{V_{DD}}{V_{GG}} = 5 \times \frac{25}{10} = 12.5 \ M\Omega$$

$$R_1 = \frac{R_G R_2}{R_2 - R_G} = \frac{5 \times 12.5}{12.5 - 5} = 8.33 \ M\Omega$$

Since

$$V_{DD} - I_D R_L - V_{DS} - I_D R_S = 0$$

$$R_L = \frac{V_{DD} - V_{DS} - I_D R_S}{I_D} = \frac{25 - 6 - 11.5}{3 \times 10^3}$$

$$= 2.5 \ k\Omega$$

We can make a check on the stability by assuming that V_{GS} shifts by 25%. Then the change in $V_{GG} - V_{GS}$ (and in I_D) would be only about 3.2%.

3.5.2   MOSFET biasing

The insulated-gate FET which operates only in the enhancement mode (refer back to page 47), has to be biased in a manner similar to that discussed above in relation to the junction-gate transistor. However, because the significant feature of the behaviour of this type of FET is that it is a normally off device, that is, the drain current is zero when the gate voltage is zero, a positive bias is necessary to establish the proper operating point. Hence the bias voltage for an enhancement FET lies between the source and drain potential levels, unlike the depletion JUGFET where it lies below or outside that range. For simple biasing, therefore, a gate resistor may be connected from the gate to the required positive level as shown in Fig. 3.17(a), though better stability of drain current will result if it is returned to the drain terminal as shown in Fig. 3.17(b). These arrangements remind us of the simple bipolar biasing discussed earlier. It must be kept in mind however that with the FET we are applying a voltage bias to a reverse biased diode, whereas with the bipolar transistor we are applying current bias to a forward biased diode.

    Biasing the form of IGFET which can operate in either the enhancement or depletion mode is often considerably simplified by

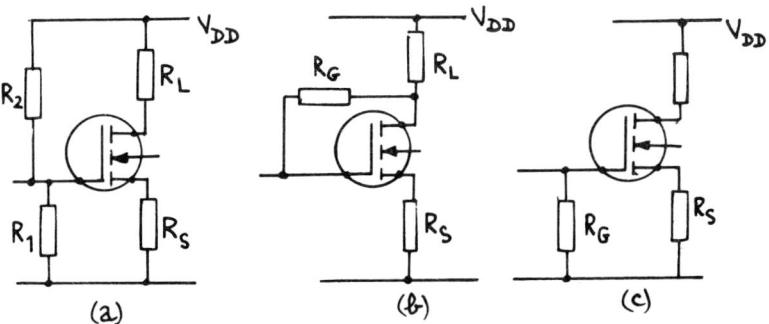

Figure 3.17

reason of the fact that a suitable operating point can be specified for $V_{GS} = 0$. The input signal excursions can then carry the device into enhancement or depletion mode as the gate moves alternately positive and negative about zero. Fig. 3.17(c) shows the circuit arrangement for zero gate bias; the gate terminal is returned to earth through a large value resistor.

3.6 HANDLING MOSFET DEVICES

Very great care is necessary in mounting and handling MOSFET devices, whether in discrete or integrated form. As explained earlier, the gate, insulating oxide layer and substrate form a capacitor in this type of transistor and if a small charge is brought into proximity with the gate terminal, a voltage will be induced across the capacitor. Since $V = q/C$, and the value of C is probably only of the order of a picofarad, the voltage appearing across the oxide dielectric will be about 10^{12} times the charge q. Hence, an extremely small charge can produce a very large potential across the dielectric and breakdown can follow. Certainly, any voltage greater than some 50 V will destroy the gate insulation. Hence a MOSFET can be destroyed simply by touching the gate terminal with a finger or the tip of a soldering iron! To prevent such damage, which would be bad enough in a single transistor but catastrophic in a number of integrated circuits, such devices are usually supplied with their connecting pins shorted together by being embedded in a conducting foam

81

or metal foil strip. When mounting MOSFETS into a board or holder, such a shorting strip should be retained in place as long as possible, and soldering iron cases should be very securely earthed.

Some MOSFETs are made with a reverse-biased diode wired internally to the gate. This diode acts as a zener if the gate voltage exceeds its breakdown potential and the gate is effectively short-circuited by the diode conduction. The very high input impedance of the MOSFET is degraded to that of the reverse biased diode and is roughly equal to that of the junction-gate FET.

PROBLEMS 3

1. Distinguish between amplitude, frequency and phase distortion.

2. Explain thermal instability as applied to a bipolar transistor.

3. What is the purpose of biasing a bipolar or field-effect transistor?

4. Explain how the circuit of Fig. 3.17 stabilizes I_C if V_{CE} changes.

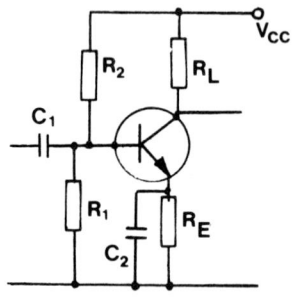

Figure 3.18

5. Sketch the constructional form and describe the principle of operation of a junction-gate FET.

6. Sketch the circuit symbols for (a) a JUGFET, (b) a depletion mode MOSFET, (c) an enhancement mode FET, and explain the symbols in terms of the operating principle of each type of transistor.

7. Why should an insulated-gate FET be carefully handled?

8. A certain transistor has α_B = 0.993 and I_{CBO} = 2 μA. What is the value of the collector current when i_B = 30 μA?

9. Explain carefully why a transistor must be biased by a method which prevents excessive shift of the d.c. operating point. What effects are likely to follow if the d.c. stabilisation is insufficient?

10. If I_C = 1.35 mA and I_E = 1.37 mA, what is I_B? If for this transistor I_{CBO} = 5 μA, what are α_B and α_E? Find I_C when the base is disconnected. Is there a value of I_E such that it is 'exactly' identical with I_C?

11. At room temperature, two transistors A and B have α_B values of 0.995 and 0.997 respectively. Also A has I_{CBO} = 3 μA and B has I_{CBO} = 8 μA. Find the α_E and I_{CEO} values for these transistors. If the collector current of A is set at 1.5 mA, find I_B. If this value of I_B is maintained constant, what will be the collector current of B when it is substituted for A?

12. What are the general rules about the voltage drop across R_E and the bleed current relative to the required I_B in a single supply potential divider biased common-emitter amplifier? Using these rules assess suitable values for the resistors shown in Fig. 3.19. Leakage current may be ignored.

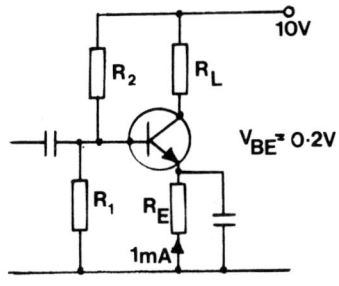

Figure 3.19

13. Using the indicated values shown on the common-emitter amplifier of Fig. 3.20, find (a) R_L, (b) the co-ordinates of the operating point. What value of R_2 is required for these co-ordinates to be obtained?

83

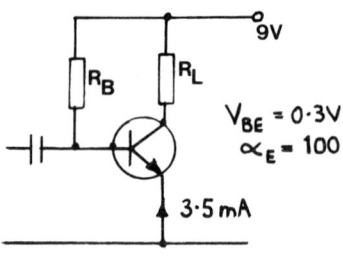

$V_{BE} = 0.3V$

$\alpha_E = 100$

3·5 mA

Figure 3.20

14. It can be shown that for a germanium transistor, the leakage current I_{CBO} is given very closely by the expression $I_o.e^{0.08}(T - T_o)$ where I_o is the value of I_{CBO} at room temperature, T is the operating temperature and T_o is room temperature, both in °C. If I_{CBO} was 10 μA at 20°C, at what temperature would it be 20 μA? What general rule of thumb can you deduce from your answer about the increase in leakage current in a germanium transistor?

15. Using the characteristics shown in Fig. 3.21, draw d.c. load lines for R_L = 2.5 kΩ, 3.75kΩ and 6 kΩ, given that V_{CC} = 15 V. Locate operating points at I_B = 20 μA for each line. What is the gradient of each line? Which load line would probably be best for an input signal which has a peak value of 20 μA?

16. What kind of collector loads are indicated by (a) a vertical load line, (b) a horizontal load line? Do you think either of these cases are realizable in practice, or closely so?

17. Refer to the circuit diagram of Fig. 3.13. At 20°C the operating point is located at I_C = 1.5 mA, V_{CE} = 6 V. I_{CBO} has a value of 1 μA at 20°C and 10 μA at 45°C. If α_B = 0.99 and is assumed constant, evaluate the shift in the operating point when the temperature rises to 45°C.

18. In a batch of transistors the lowest gain transistor has α_B = 0.96 and the highest has α_B = 0.98. Also, the low gain transistor has I_{CBO} = 5 μA and the high gain transistor has I_{CBO} = 20 μA. If I_E is fixed at 1 mA, find the limits in collector current.

Suppose in these transistors that α_B increases by 2 parts in 10^4 per $^{\circ}$C rise and that I_{CBO} doubles for every 10°C increase. If the low gain unit is used at room temperature (20°C) but the high gain unit is used at 50°C find the new limits in collector current.

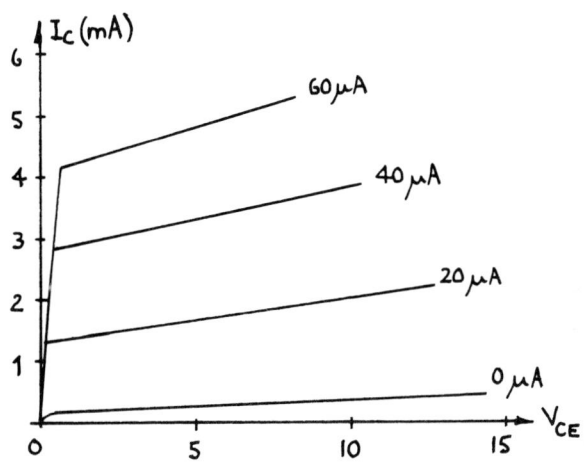

Figure 3.21

4 Large signal amplifiers

4.1 INTRODUCTION

Although the integrated power amplifier is now commonplace in a
variety of power ratings, power amplification is still very widely
the province of the discrete transistor. It seems logical then to
deal with power amplifiers in discrete transistor circuitry at this
stage as a direct follow-on from the fundamental considerations of
the previous two chapters. Integrated circuit forms will be developed from this approach and these will in turn serve as an introduction to the small-signal amplifier where integrated circuits have,
in the general sense, superseded the discrete transistor circuit
systems of a few years ago.

4.2 POWER AMPLIFIERS

A large-signal or power amplifier is designed, by definition, to
supply power to an output load. As this power may range from a few
watts to many hundreds of watts the main design consideration must be
the elimination of power wastage in the amplifier. This means that
efficiency is the parameter of primary importance. But there are
other factors which are peculiar to power amplifiers; of these, distortion and heat dissipation are the next most important. Distortion
becomes a problem because, unlike small-signal amplifiers, power
transistor parameters vary appreciably over the input cycle and the
requirements of high power mean that operation runs into the non-
linear regions of the device characteristics. Heat dissipation also
becomes a problem because a relatively large proportion of the d.c.
power input is wasted as heat within the transistor(s) and care has

to be taken that this dissipation does not exceed the maximum permissible junction temperature. We shall discuss in this chapter these special difficulties and determine how best power amplifiers can be designed to operate without introducing excessive distortion and without exceeding the voltage, current, power and temperature limitations of the active devices.

4.2.1   Operating boundaries

Power amplifiers are best dealt with by graphical analysis. A circuit of a simple Class-A common-emitter power amplifier is given in Fig. 4.1, and a typical set of collector characteristics in Fig. 4.2. Using this latter diagram, we can determine how the operating line and the load line are positioned, and derive expressions for the power output and stage efficiency.

The maximum power that a given transistor can dissipate under

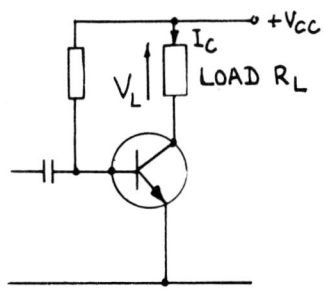

Figure 4.1

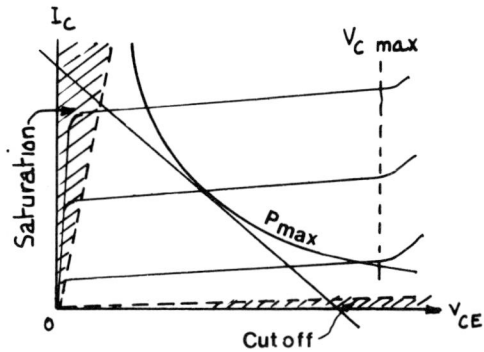

Figure 4.2

stated ambient conditions can be shown on the output characteristics
by drawing the locus of points satisfying $V_{CE}I_C = P_{C\ max}$, where
$P_{C\ max}$ is the maximum collector dissipation established by the manu-
facturer. This leads to a rectangular hyperbola as shown in Fig.
4.2. Continuous operation outside this boundary is not permissible.
There are other boundaries which must be observed also if excessive
non-linearity is to be avoided. The shaded areas adjacent to the co-
ordinate axes indicate those regions in which (a) collector current
is approaching zero (the 'cut-off' region) and a further decrease
in the signal input current (i_B) does not produce a corresponding
decrease in output current; (b) collector voltage is approaching
zero (the 'saturation' region) and a further increase in input signal
current does not produce a corresponding increase in output current.
There is also a limit to the greatest collector voltage which can be
tolerated. Operation beyond this limit will create the avalanche
breakdown effect and the transistor may be damaged.

The biasing of the stage must be such that operation takes place
within the indicated boundaries. Further, within the limitations
imposed by these boundaries, we require the maximum useful power
output. The load line, when drawn, should therefore lie below the
maximum dissipation curve and its gradient should be such that the
opposing requirements of large power output and low distortion are
properly proportioned.

The load line does not have to touch the hyperbola but it is
clear that the maximum voltage and current excursions are increased
as it approaches the hyperbola as a tangent. There are, of course,
any number of possible tangents available and Fig. 4.3 shows two

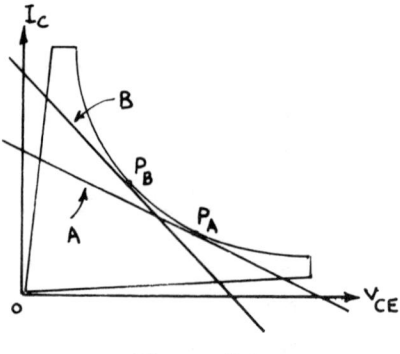

Figure 4.3

possibilities. Load line A is of small gradient i.e. the load resistance is large. This permits a large voltage swing but restricts the current variation. Load line B has a large gradient which permits a large current swing but restricts the voltage variation. There is no particular optimum position for the load line but it should be tangential to the power hyperbola and the operating point should be located on the point of contact. Beyond that, such other factors as load resistance (which may or may not be adjustable) and the available supply voltage must come into the picture.

4.2.2   The A.C. load line

We have so far assumed that the gradient of a load line depends only upon the value of the load resistance and that the operating conditions are purely d.c. It is not often that the load line on which the amplifier actually operates is the one drawn for such d.c. conditions. The reason for this is that the load into which the transistor works is different for a.c. (signal) and d.c. (static) conditions.

Figure 4.4 shows a practical amplifier which includes an emitter resistor R_E. This resistor is shunted with capacitor C whose reactance at the lowest signal frequency is negligibly small. It is therefore an a.c. short-circuit so that as far as the signal is concerned R_E is effectively removed. Hence the effective load on the transistor is R_L only and the equation of the load line (for a.c. or dynamic conditions) is $V_{CC} = I_C R_L + V_{CE}$.

For d.c. conditions on the other hand, C is an open-circuit so that the d.c. load line and the operating point are determined from the equivalent d.c. circuit where the load on the transistor is

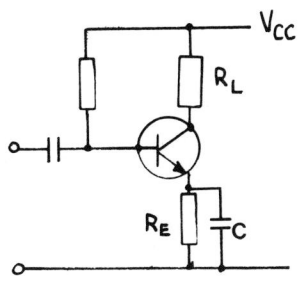

Figure 4.4

$(R_L + R_E)$. The equation of the d.c. (or static) load line is therefore $V_{CC} = I_C(R_L + R_E) + V_{CE}$. Taking the operating point to be about
the centre of the load line, $V_{CC} \simeq 2V_{CE}$ and then $(R_L + R_E) \simeq V_{CC}/I_C$.
Since the gradient of the a.c. load line is $-1/R_L$ while that of the
d.c. load line is $-1/(R_L + R_E)$, the a.c. load line must be 'steeper'
than the d.c. load line. Both lines are shown on the characteristics
of Fig. 4.5.

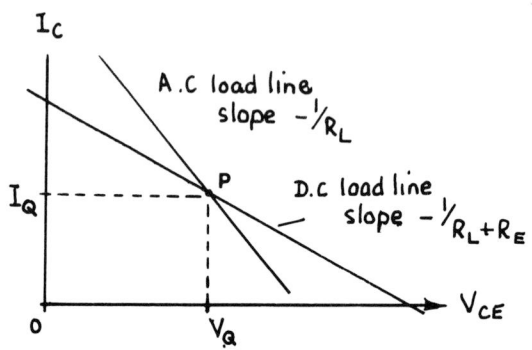

Figure 4.5

To establish the proper operating point now, the d.c. load line
is drawn first and a suitable point P is selected. The a.c. load
line is then drawn passing through P and having a gradient equal to
$-1/R_L$. It is particularly important to note that the a.c. line MUST
pass through P and is not established (as is the d.c. load line) by
the points of intersection on the respective axes. Where R_E is small
in comparison with R_L, the a.c. and d.c. load lines may for convenience be considered identical.

Example 4.1   The output characteristics for the transistor used in
the circuit of Fig. 4.6(a) are shown at (b). If the required base
bias is 10 mA, draw the d.c. and a.c. load lines on the given characteristics.

From the diagram $R_L = 600 \, \Omega$. This load is effectively shunted
by R_1 under a.c. conditions since capacitor C is an a.c. short-
circuit and V_{CC} is a steady d.c. source having a negligible impedance to earth. Therefore the upper end of R_L and the lower end
of R_1 are at the same a.c. potential; hence R_L and R_1 are in parallel
to the signal frequency. The a.c. load is therefore $R_L \| R_1 = 450 \, \Omega$.

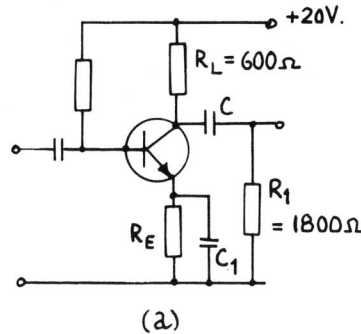

(a)

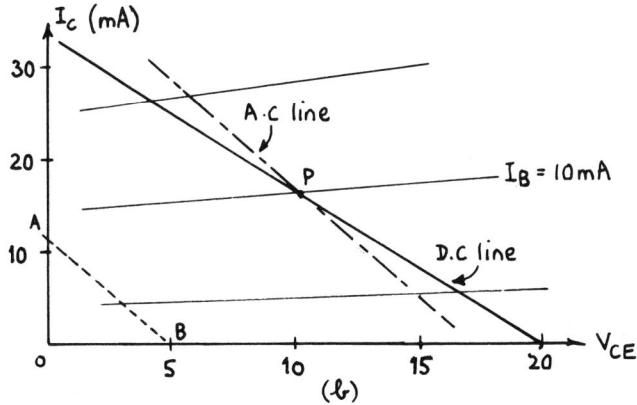

(b)

Figure 4.6

Under d.c. conditions, capacitor C is an open-circuit, hence R_1 has no effect on R_L and the d.c. load is 600 Ω.

The d.c. load line is constructed first in the usual manner; it passes through V_{CC} = 20 V on the V_{CE} axis and cuts the I_C axis at V_{CC}/R_L = 33.3 mA. Its gradient is clearly -1/600 S. The operating point is located at the intersection of the load line and the I_B = 10 mA characteristic. The a.c. load line has a gradient of -1/450 S and passes through P. To draw this we use a 'dummy' line shown as A-B. This is simply a line having the required gradient drawn between any two convenient points on the axes. The required a.c. load line is now drawn parallel to A-B and passing through P.

As an exercise, you should be able to think of an alternative way of positioning the a.c. load line.

4.3   THE TRANSFORMER-COUPLED AMPLIFIER

Although with the advent of integrated power amplifiers the transformer-coupled amplifier has largely vanished from the audio scene,
it has many applications in industrial control systems and forms a
useful introduction to those circuit configurations in which no
transformers are used. With transformer coupling, shown in essence
in Fig. 4.7, several useful features emerge:

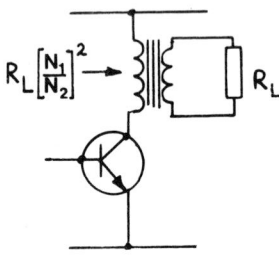

Figure 4.7

1.   Only a changing flux induces voltage, so direct currents are
isolated from circuits connected to the secondary terminals.
2.   The transformer is an impedance matching device and facilitates
the matching of widely differing load impedances to the load requirements of the transistor(s).
3.   Since the 'd.c.' resistance of the primary coil is low, the necessary supply voltage to the amplifier can be reduced, V_{CE} being
practically to V_{CC}.
4.   The efficiency of a transformer is high.

Since the changing flux in the transformer core resulting from
the a.c. signal variations in the primary winding induces voltages
in the windings proportional to the number of turns, then V_2/V_1 =
N_2/N_1. The current ratio must be the inverse of this voltage ratio
since (assuming no loss) the power output must equal to the power
input.   Then $I_2/I_1 = N_1/N_2$. If an impedance Z_L is connected to the
secondary, the impedance seen at the primary is

$$Z'_L = \frac{V_1}{I_1} = \left[\frac{N_1}{N_2}\right]^2 \cdot \frac{V_2}{I_2} = \left[\frac{N_1}{N_2}\right]^2 \cdot Z_L$$

Therefore (4.1)

$$Z'_L = n^2 Z_L$$

where n is the transformation ratio N_1/N_2.

Figure 4.8 shows a simplified circuit diagram of a single ended transformer-coupled Class A output power amplifier. Bias is obtained by way of potential divider R_1 and R_2 and an emitter resistor is

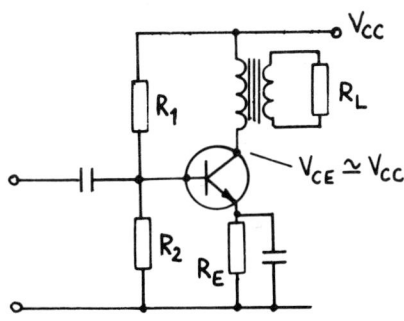

Figure 4.8

included for thermal stability. Assuming that R_E is small and that the d.c. resistance of the transformer primary winding is negligible, then the d.c. power supplied $P_{dc} = V_{CC} \cdot I_Q$ where I_Q is the steady or quiescent collector current flowing at the selected operating point. But the power dissipated at the collector is the product of the collector voltage and the quiescent current, so $P_C = V_Q \cdot I_Q$. This is the same as the d.c. power supplied since the quiescent collector voltage V_Q is now equal to V_{CC}. Hence the whole of the power supplied is dissipated as heat at the collector junction.

When a sinusoidal signal voltage V_i is applied to the input terminals, it will vary about the mean base bias as shown in Fig. 4.9(a) and the collector current I_c will vary in phase with V_i about the quiescent value I_Q, see Fig. 4.9(b). To this alternating current the transformer primary resistance is $n^2 R_L$, hence the collector voltage V_c will vary in antiphase to the collector current about the quiescent value V_Q (= V_{CC}), as shown in Fig. 4.9(c).

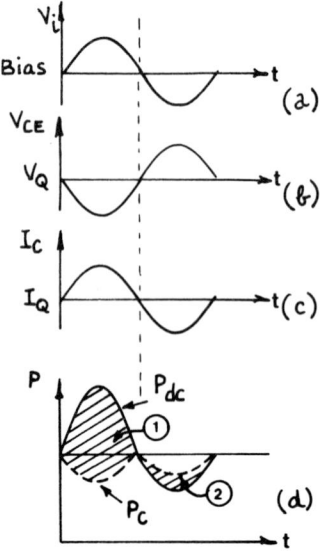

Figure 4.9

Now the power supplied by the d.c. source is the product of V_{CC} and I_Q, hence the curve of d.c. power supplied is the product of curve (b) and the constant V_{CC}. This curve is shown in the full line of Fig. 4.9(d). The power dissipated at the collector is the product $I_Q V_Q$, hence the curve of power dissipated is the product of curves (b) and (c). Since these are phase opposed, their product will be wholly negative and the resultant wave this time is shown in broken line in diagram (d).

The area under a power-time graph represents energy in joules, so by shading in the areas between the two curves the energy representation of what is happening during each half-cycle of input signal between the power supplied (P_{dc}) and the power dissipated at the collector (P_C) can be demonstrated. As the areas are not equal, some of the energy supplied must go elsewhere other than as dissipation and this must obviously be as energy delivered to the load. During the positive half-cycle of input, energy given to the load is positive i.e. energy given by the source is GREATER than the collector dissipation. During the negative half-cycle the energy given by the source is LESS than the collector dissipation. The useful power output into the load is thus represented by area 1 minus area 2, that is $P_L = P_{dc} - P_C$. Since P_{dc} is constant, THE COLLECTOR DISSIPATION IS GREATEST WHEN THE INPUT SIGNAL IS ZERO and $P_L = 0$. All transistors (and valves) have a figure quoted for the maximum power they can dissipate; we have covered this aspect in Section 4.2.1. Now

$$\text{Efficiency } \eta = \frac{\text{mean power supplied to the load}}{\text{mean power supplied by the source}}$$

$$= \frac{\text{power in the load}}{\text{power in the load + total dissipation}}$$

$$= \frac{P_L}{P_L + P_C}$$

Strictly, the total dissipation must include the small resistive loss occurring within the transformer primary, but this is usually small and we shall normally assume that the collector dissipation represents the only power wastage.

Example 4.2   A Class-A power amplifier draws a quiescent collector current of 750 mA from a 25 V supply and delivers a signal power of 5.8 W to the collector load. Calculate (a) the stage efficiency, (b) the collector dissipation.

$$P_{dc} = V_{CC}I_Q = 25 \times 750 \times 10^{-3}$$

$$= 18.75 \text{ W}$$

(a) $\eta = \dfrac{P_L}{P_{dc}} = \dfrac{5.8}{18.75} = 0.31 \ (31\%)$

(b) $P_C = P_{dc} - P_L = 18.75 - 5.8$

$$= 12.95 \text{ W}$$

4.3.1   Theoretical efficiency

The theoretical limit for efficiency for a Class-A amplifier can be determined by considering the characteristic shown in Fig. 4.10. Let the signal excursions along the load line be QS, centred at the operating point P. The power in the load R_L is clearly

$$P_L = \frac{(\hat{I} - I_Q)^2 R'_L}{2}$$

while the power supplied is $V_{CC}I_Q$.

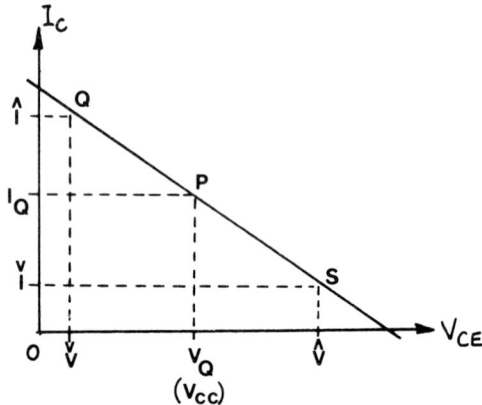

Figure 4.10

Hence

$$\eta = \frac{P_L}{P_{dc}} = \frac{(\hat{I} - I_Q)^2 R'_L}{2V_{CC}I_Q} \tag{4.2}$$

Assuming that the full length of the load line is utilised, $\check{V} = 0$, $\hat{V} = 2V_{CC}$, $\hat{I} = 2I_Q$ and $R'_L = V_{CC}/I_Q$. Substituting these maximum values into equation (4.2) gives

$$\eta = \frac{(2I_Q - I_Q)^2 \cdot V_{CC}/I_Q}{2V_{CC}I_Q} = 0.5 \text{ or } 50\%$$

This is the maximum theoretical efficiency. Since only about 90% of the load line can be utilised at maximum input signal drive, the actual efficiency of transistor amplifiers is about 40-45%. The AVERAGE efficiency resulting from a variable amplitude signal is usually very much lower than this.

Example 4.3   The operating line for a power amplifier having a 15Ω directly connected in the collector lead is shown in Fig. 4.11. The output voltage swings between 2 V and 28 V and the output current swings between 0.1 A and 1.9 A. Find the stage effieicncy and

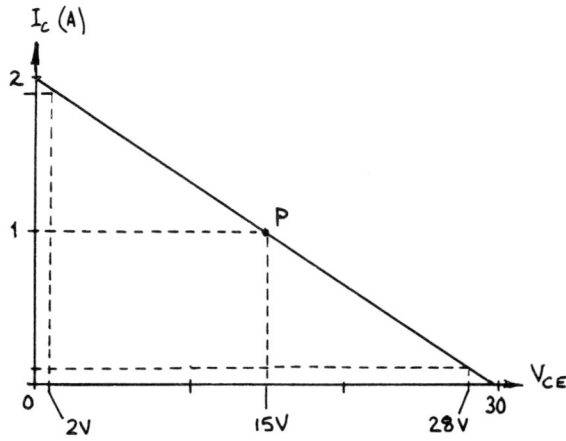

Figure 4.11

compare this with that obtained from a transformer-coupled stage
operating along the SAME load line.

Here clearly V_Q = 15 V, I_Q = 1 A and V_{CC} = 30 V. The signal
power in the load is the product of the rms voltage and current
acting at the load, that is

$$P_L = \frac{(28 - 2)}{2\sqrt{2}} \frac{(1.9 - 0.1)}{2\sqrt{2}} = 5.85 \text{ W}$$

The power input with 15 Ω in the collector is P_{dc} = V_{CC} = 30 W.
Hence the efficiency is 5.85/30 = 0.195 or 19.5%.

When the load is transformer coupled, V_{CE} will swing as high
above V_{CC} as it swings below it. Then the required $V_{CC} = \frac{28 + 2}{2} =$
15 V and the signal power in the load will be the same as before.
The input power, however, will now be P_{dc} = $V_{CC}I_Q$ = 15 W, and the
collector efficiency will be 5.85/15 = 0.39 or 39%. twice as great
as before.

4.4   PUSH-PULL OPERATION
By using transistors in what is known as a push-pull arrangement, a
greater power output than is possible with a single-ended stage becomes possible. Other advantages are also obtained such as lower
harmonic distortion and, by a proper adjustment of the operating
point, a much greater efficiency. A push-pull amplifier arrangement

is shown in Fig. 4.12. Transformer coupling is used to add the balanced outputs of the two transistors into the single load R_L, and is also used to provide balanced anti-phase input signals to the two bases.

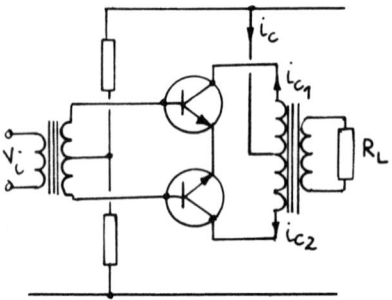

Figure 4.12

4.4.1   Class-A push-pull

By biasing each transistor to the centre of the load line, the amplifier operates in Class-A; when the transistors are driven by input signals of opposite polarity the outputs are combined in the transformer to give an increased power in the load. This power is approximately twice that obtainable from a single transistor under identical operating conditions. However, because the normal quiescent collector current is also doubled, the collector efficiency remains the same. Fig. 4.13(a) and (b) show respectively the d.c. and a.c.

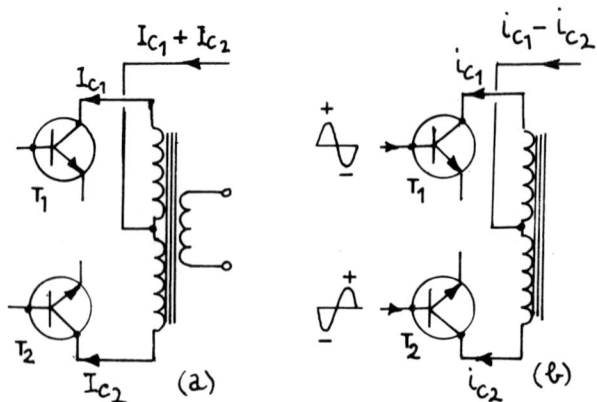

Figure 4.13

conditions acting in the circuit. The d.c. collector currents flow
in the transformer primary sections in opposite directions, hence
there is no resultant d.c. magnetisation of the core. This means
that a smaller transformer can be used and a high primary inductance
maintained. The signal currents, on the other hand, ADD in the
transformer primary since the collector current of T_1 will be decreasing at the same time as that of T_2 will be increasing, and conversely;
hence the name 'push-pull'. From the point of view of the V_{CC} supply
line, these a.c. variations cancel out, hence no signal currents get
into the supply line and the need for heavy decoupling is reduced.
Further, as a result of the balanced arrangement, even harmonics, if
present, will also be cancelled out. Figure 4.14 illustrates the
action of Class-A push-pull with respect to the dynamic transfer
characteristics of the two transistors.

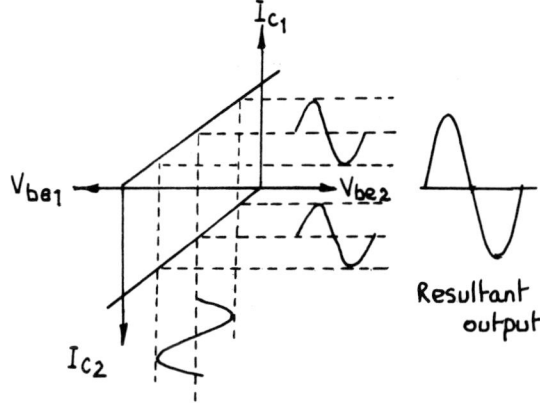

Figure 4.14

Care has to be taken with push-pull circuits to get the transformer matching ratio right. In Class-A operation the load current
is doubled over that of a single ended stage but the voltage is unchanged. If each transistor is to see the same effective load as it
did in the single ended case, the value of R_L must be halved, assuming that we use the equivalent of two single ended primaries in
series i.e. an n : 1 transformer becomes an (n+n) : 1. If a particular value of load is to be driven, the turns ratio must, of course,
be adjusted accordingly.

4.4.2   Class-B push-pull

Where the push-pull circuit becomes advantageous is in the fact that
if the operating points of the individual transistors are moved towards the lower end of the respective load lines, that is, towards
collector current cut-off, the quiescent power requirement is drastically reduced and the efficiency thereby increased.   This type of
operation is known as Class-AB or Class-B,   Class-B proper referring
to the case where the operating point is placed precisely at cut-off,
see Fig. 4.15.   The whole of the load line then becomes available for
each half-cycle of the input signal.   Under this condition $I_Q = 0$ and
$V_Q = V_{CC}$.   The load line then connects the operating point P with
$I_C = I_C$, $V_{CE} = 0$ and $R'_L = V_{CC}/I_C$.

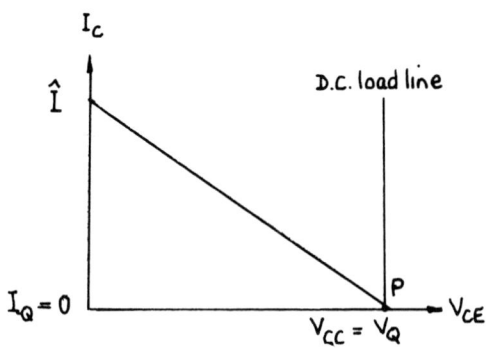

Figure 4.15

A sinusoidal input now simultaneously drives one transistor on
and the other beyond cut-off so that the devices conduct on alternate
half-cycles to provide consecutive half-wave current pulses in the
primary winding of the transformer.   As in the case of Class-A
operation, the transformer combines these half-wave pulsations into
the required full-wave output across the load resistance.   Figure
4.16 illustrates this.

Although each transistor now provides half the power output, the
load presented to the conducting transistor must be such that the
TOTAL power is dissipated over its operative half cycle.   Figure 4.17
illustrates the action of Class-B push-pull with respect to the
dynamic transfer characteristics of the transistors.   Compare this
diagram with that given for Class-A earlier.   Make a point of

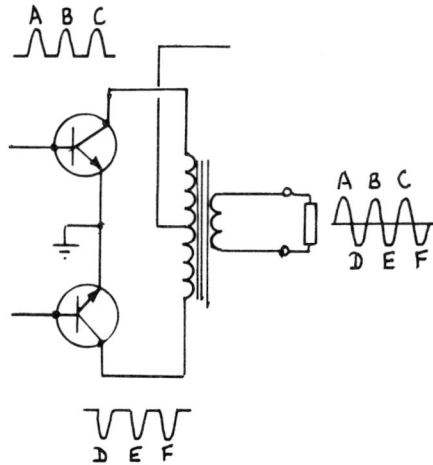

Figure 4.16

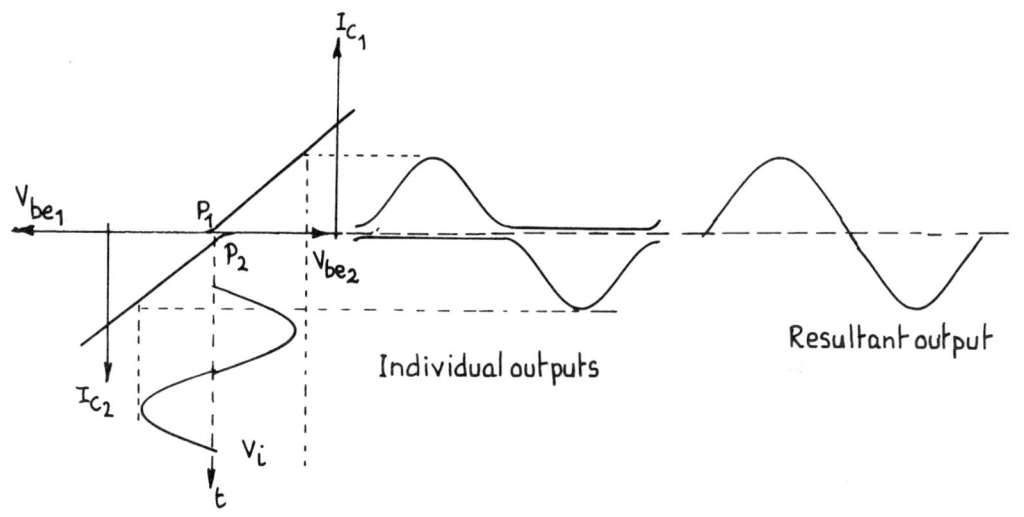

Figure 4.17

noticing that the operating points in Fig. 4.17 are not positioned
exactly at collector current cut-off but are placed so that the COM-
POSITE transfer characteristic obtained from the summation of the individual characteristics is a straight line.

4.4.3   Theoretical efficiency

In analysing a push-pull arrangement, it is customary to treat only
one of the transistors, since each is operating into an identical load
at identical levels.   The efficiency can therefore be determined by
considering only one device working with a half-sinewave signal.   Both
the load current and load voltage are half-sinewaves but V_{CC} is of
course constant.   We can represent the power conditions in the circuit by the waveforms shown in Fig. 4.18.   For the d.c. supply power,
this is simply the product of V_{CC} and the 'mean' load current; during each half-cycle this current is $\hat{I}/\pi$, hence the power supplied,
as the figure illustrates, is

$$P_{dc} = V_{CC} \cdot \frac{\hat{I}}{\pi}$$

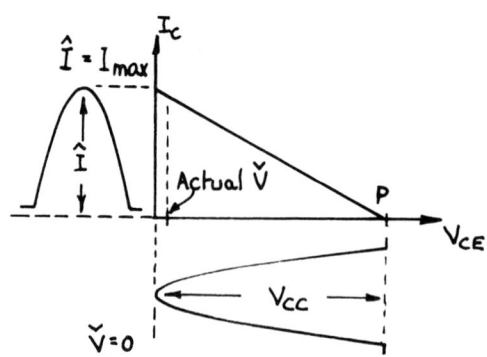

Figure 4.18

For the signal power in the load, this is the product of load
voltage V_L and load current I_L.   Then the power delivered by a stage
that is driven through its total range of saturation to cut-off is
given by

$$P_L = \frac{1}{2}\left[\frac{\hat{V}}{\sqrt{2}} \cdot \frac{\hat{I}}{\sqrt{2}}\right] = \frac{\hat{V}.\hat{I}}{4}$$

Notice that the ½ factor is used because we are dealing with half-
wave pulses.   The maximum efficiency for each transistor and therefore for the amplifier as a whole is then

$$\frac{P_L}{P_{dc}} = \frac{\hat{V}}{4V_{CC}} = \frac{\pi}{4} = 0.785 \quad \text{or} \quad 78.5\%$$

since, if the whole of the load line is utilized, $\hat{V} = V_{CC}$. The corresponding peak powers are then

$$\hat{P}_L = V_{CC}^2/4R_L' \quad \text{and} \quad \hat{P}_{dc} = V_{CC}^2/\pi R_L'$$

since $\hat{I} = V_{CC}/R_L'$.

In practice the load line cannot be fully utilised, particularly at the saturation end as shown in Fig. 4.18. Let the load voltage swing from V_{CC} to $\check{V}$ when the load current rises from zero to $\hat{I}$. The mean a.c. power in the load is then

$$P_L = \frac{1}{2}\left[\frac{\hat{I}}{\sqrt{2}}\frac{(\check{V}_{CC} - \check{V})}{\sqrt{2}}\right] = \hat{I}\frac{(V_{CC} - \check{V})}{4}$$

Also, as before, $P_{dc} = V_{CC}\cdot\dfrac{\hat{I}}{\pi}$

Therefore

$$\eta = \hat{I}\frac{(V_{CC} - \check{V})\pi}{4V_{CC}\hat{I}} = \frac{\pi}{4}\left[1 - \frac{\check{V}}{V_{CC}}\right]$$

For a finite $\check{V}$, this is always less than $\pi/4$.

Example 4.4   Express the collector dissipation directly in terms of signal power in the load and the efficiency.
We have

$$P_c = P_{dc} - P_L$$

$$= P_L\left[\frac{P_{dc}}{P_L} - 1\right]$$

Therefore

$$P_C = P_L\left[\frac{1}{\eta} - 1\right] \tag{4.3}$$

Example 4.5   Two identical transistors are used in a Class-B push-pull amplifier. They are required to deliver a continuous output power of 20 W and work from a 12 V supply rail. If the collector voltage swing is sinusoidal with a peak of 10 V, find (a) the stage efficiency, (b) the collector dissipation.

Power output per transistor = 10 W, collector voltage swing = 10 V peak. The minimum collector voltage $\check{V}$ = 2 V.

$$\eta = \frac{\pi}{4}\left[1 - \frac{2}{12}\right] = \frac{5\pi}{24} \approx 0.65 \quad \text{or} \quad 65\%$$

Dissipation $P_C = P_L\left[\frac{1}{\eta} - 1\right] = 10\left[\frac{1}{0.65} - 1\right]$

$$= 5.38 \text{ W}$$

4.4.4   Power relationships

With no signal input both P_{dc} and collector dissipation are zero (or very closely so). P_{dc} is a maximum at full drive since the mean collector current is then a maximum. The maximum dissipation, however, does NOT occur at full drive as the mathematics will show.

$$P_C = P_{dc} - P_L$$

$$= V_{CC}\cdot\frac{\hat{I}}{\pi} - \frac{\hat{I}^2 R_L'}{4} \qquad\qquad (4.4)$$

since $P_L = \frac{1}{2}[\frac{\hat{I}}{\sqrt{2}}]^2 R_L'$. Differentiating for a maximum

$$\frac{dP_C}{d\hat{I}} = \frac{V_{CC}}{\pi} - \frac{\hat{I}.R_L'}{2} = 0$$

Hence $\hat{I} = \frac{2V_{CC}}{\pi R_L'}$ compared with $\hat{I} = \frac{V_{CC}}{R_L'}$ at full drive. The transistor heat is therefore greatest when the signal amplitude is $2/\pi$ (≈ 0.64) of that giving maximum power.

Substituting this result into (4.4) then yields the required peak power:

$$\hat{P}_C = \frac{V_{CC}}{\pi}\left[\frac{2V_{CC}}{\pi R_L'}\right] - \frac{R_L'}{4}\left[\frac{2V_{CC}}{\pi R_L'}\right]^2$$

$$= \frac{2V_{CC}^2}{\pi^2 R_L'} - \frac{V_{CC}^2}{\pi^2 R_L'} = \frac{V_{CC}^2}{\pi^2 R_L'}$$

This value must clearly not exceed $P_{C\ max}$.

The advantage of Class-B over Class-A working is also illustrated by comparison of the ratio

$$\frac{\text{peak signal output}}{\text{maximum dissipation}} = \frac{\hat{P}_L}{P_C}$$

for the two arrangements. For ideal Class-A working, the maximum output is one-half of the power input but the maximum dissipation which occurs with no drive is equal to the power input. Hence the ratio is 0.5.

For ideal Class-B working there is a considerable improvement on this for

$$\frac{P_L}{P_C} = \frac{V_{CC}^2 \pi}{4R_L'} \cdot \frac{R_L'}{V_{CC}^2} = \frac{\pi^2}{4} \simeq 2.5$$

This means that if the power dissipation capability, for example, of a transistor used in either form of amplifier is 10 W, the maximum power output from Class-A working is 5 W while that from the Class-B amplifier is about 25 W.

Example 4.6   A Class-B output stage has two transistors each rated at 10 W maximum dissipation. Find the maximum power output for a sinusoidal input and an actual efficiency of 65%. What would be the power output for an efficiency of 50%?

$$P_{dc} = P_L + P_C = P_L + 10$$

When

$$\eta = 0.65 \qquad\qquad \frac{P_L}{P_{dc}} = \frac{P_L}{P_L + 10} = 0.65$$

Therefore

$$0.65P_L + 6.5 = P_L$$

$$P_L = 18.6 \text{ W}$$

If $\eta = 0.5$, then $\dfrac{P_L}{P_L + 10} = 0.5$

$$P_L = 10 \text{ W}$$

This is a decrease of almost 100% for a relatively small drop in the efficiency.

4.5   TRANSFORMERLESS CIRCUITS

Transformers are of necessity large components and the necessary
characteristics that they must have to perform as efficient output
transformers operating over a wide frequency range make them expensive as well as bulky. With discrete transistor circuitry it is
obviously advantageous if the transformer can be eliminated; with
integrated circuits it is clearly necessary to have output circuit
arrangements which do away with the transformer altogether. Two
basic arrangements are shown in Fig. 4.19(a) and (b). Circuit (a)
is essentially a bridge system with the load R_L connected into the
centre arm. Assuming the transistors are biased to cut-off in Class-
B, they switch on and off alternately on each input cycle. The resulting collector currents consequently act in opposite direction in
the load (as they did in the transformer primary earlier) and recon-
stitute the complete wave across the load resistance. Antiphase
input signals are still required at the base inputs.

The circuit of Fig. 4.19(b) is a modification which eliminates
the need for antiphase inputs. Here an n-p-n and a p-n-p transistor
pair are used in what is known as a 'complementary' circuit and because of the symmetrical layout the system is called 'complementary
symmetry'. What happens now is that the input signal drives one
transistor off and the other on over each half-cycle, so doing away
with the need for antiphase inputs. When the signal input is zero

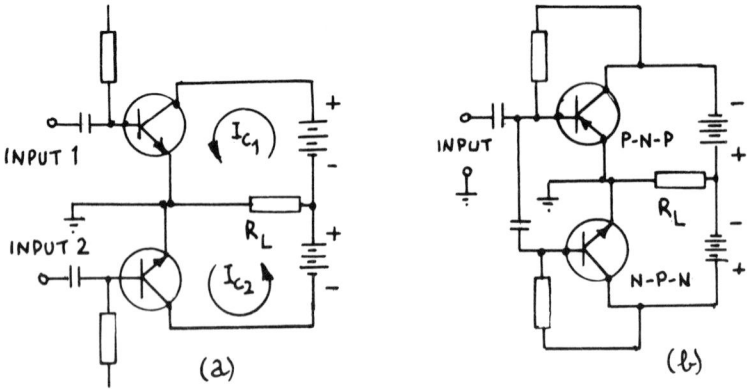

Figure 4.19

the two transistor currents are equal (or nearly so) and no direct current flows through the load.

These amplifier forms, like any push-pull system, can be operated Class-A although Class-B is most usual because of the high efficiency obtainable. It is not difficult to see limitations in the basic complementary circuit of Fig. 4.19(b). The supply is split into two parts and does not have an earth point. The output signal appears between the supply and earth as a consequence, hence the supply cannot be used for other sections of the circuit preceding the power amplifier stage. Further, the shunt capacitance existing between the supply and earth will limit the high frequency response of the amplifier.

A more practical form of the circuit is shown in Fig. 4.20. The split power supply is avoided by feeding the load through capacitor C_3 and the forward bias for the transistors is provided by the two forward-biased diodes D_1 and D_2. These diodes ensure that the voltage on the base of T_2 is slightly more positive than the voltage on the base of T_3; similarly, the voltage on the base of T_3 is slightly more negative than the voltage on its emitter. T_2 and T_3 then pass the same quiescent current which is a few milliamperes above cut-off. The low dynamic resistance of the diodes as well as their negative temperature coefficients improve the thermal stability of the circuit over that obtainable if a small resistance was used in place of the

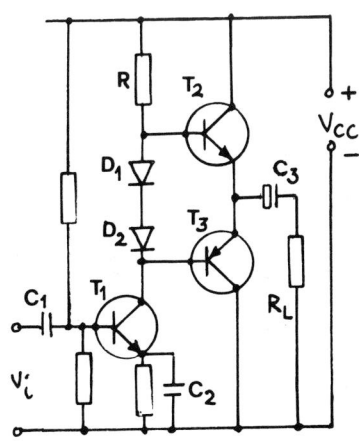

Figure 4.20

diodes. Small resistances are, however, often wired in either series or shunt with the diodes for precise adjustment of the operating point, and the diodes themselves are placed in close thermal contact with T_2 and T_3.

The driver transistor T_1 is directly coupled to the output stage. When the input signal v_i is positive, the collector of T_1 goes negative so that T_2 is cut off and T_3 conducts. When T_1 saturates at the peak of the input signal, its collector potential is above the earth line by the voltage drop across R_E plus the saturation voltage of T_1 (about 0.2 V); the potential at the emitter end of the load is then this potential plus the V_{BE} drop in transistor T_3. Nothing can be done about the saturation voltage of T_1 or the V_{BE} drop in T_3; R_E however, should be small so that the $I_E R_E$ drop is small. The output voltage can then be considered to be close to earth potential.

When v_i goes negative the collector of T_1 goes positive so that T_3 cuts off and T_2 conducts. The emitter end of the load then rises towards V_{CC}. It cannot reach this level because the collector current of T_1 and the base current of T_2 must flow through the load R, hence at the peak negative input swing T_1 is cut off and the maximum-signal base current of T_2 must flow through R. Hence the positive load potential is V_{CC} minus the voltage $i_B R$ plus V_{BE} of transistor T_2. Nothing can be done about V_{BE} in T_2 but the voltage drop in R should be small. The output voltage swing is then the greatest possible.

Circuits of this form clearly lend themselves to fabrication as integrated circuits, though there is a limit to the power output obtainable from such devices which is determined by the maximum internal power dissipation that the chip is able to handle. For very large powers, external discrete transistors have to be used, with the integrated circuit containing the small-signal and driver transistors. Where powers of up to some 20 W are needed, integrated circuits contain the output transistors and an associated heat sink (of which more in a very short while) which is brought out of the casing in the form of two thickened tabs which can be soldered to an area of the printed board on which the unit is attached or a finned aluminium radiating surface is attached to the case itself.

A circuit form commonly found in integrated circuits is shown in

Fig. 4.21. Here the collector currents of the output transistors T_3 and T_4 flow through R_1 and R_2 and the voltages developed across these resistors act to stabilize the operating point. Reading round the diode-transistor loop, we have

$$V_{BE1} + V_{BE2} + I_C(R_1 + R_2) \simeq V_{D1} + V_{D2}$$

and

$$I_C \simeq \frac{V_{D1} + VD_2 - (V_{BE1} + V_{BE2})}{R_1 + R_2}$$

where I_C $(= I_Q)$ is the quiescent collector current of T_3 and T_4. The emitter currents of T_1 and T_2 also flow through R_1 and R_2 but these may be neglected in comparison with the much larger collector current. Again, since the thermal coefficients of the diodes match those of the transistors, thermal stability is ensured by placing the diodes in close proximity to T_3 and T_4. The circuit configuration is emitter-follower, and the theory of this will be dealt with in Chapter 7.

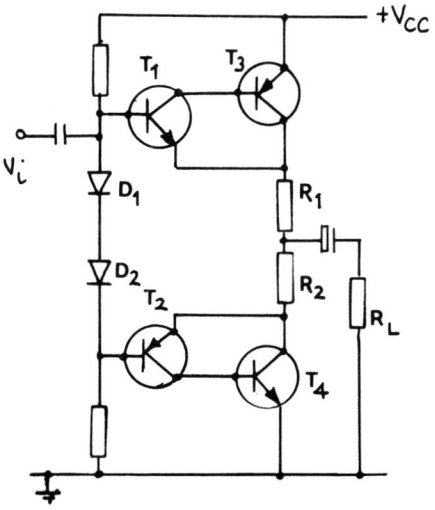

Figure 4.21

Integrated power amplifiers contain the pre-amplifiers and driver stages in addition to the output stage and a number of external components must be wired to the appropriate terminals of the

chip. These additional components are usually those involved with
decoupling, volume and tone control and negative feedback. A typical
circuit with these external components shown is illustrated in Fig.
4.22. This circuit will develop 5 W into a load resistance of 8 Ω
with a supply voltage of 12 V.

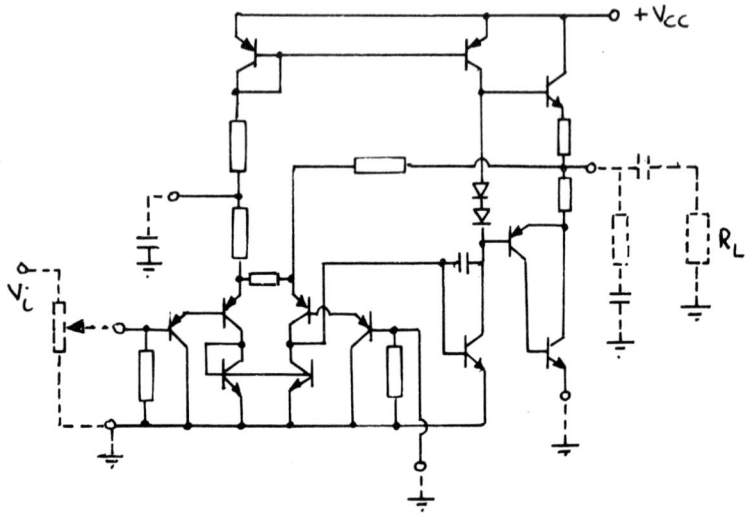

Figure 4.22

4.6   DISTORTION

Power amplifiers are what might be called 'natural' sources of distortion in so far as they are required to be driven from close to
cut-off on one half-cycle to near saturation on the next half-cycle,
that is, they are operated over a significant range of their characteristics. The parameter variations which are of importance when
large signal excursions are concerned are (a) the input characteristic which relates V_{BE} to I_B, and (b) the current gain which is
dependent upon the collector current.

The input characteristic is fundamentally that of a forward-
biased diode, hence it follows an exponential law and the input resistance varies with the signal amplitude. A typical variation is
shown in Fig. 4.23(a). The current gain is dependent upon collector
current and a typical characteristic is shown in Fig. 4.23(b). If
we imagine a stage biased at 1 A and handling a signal that ranges

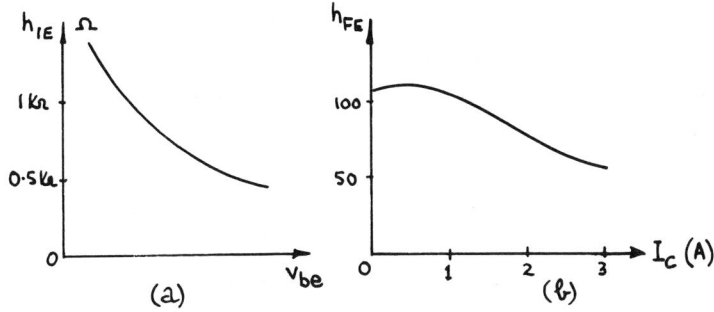

Figure 4.23

from about zero to 2 A about that point, the positive excursions would get considerably less amplification than the negative excursions.

Fortunately, these two non-linearities tend to cancel out and will result, when combined, in an overall transfer characteristic such as that shown in Fig. 4.24. However, as the figure depicts, when a signal voltage which is assumed to be a pure sinusoid is applied to the transistor input, the output waveform is distorted by being flattened on the positive peak and sharpened on the negative peak. Consequently it contains frequency components (harmonics) which were not present in the input signal.

Harmonic distortion is analysed by assuming the driving source to be sinusoidal and then finding the amplitude coefficients of the first few terms of the Fourier series representing the distorted wave. Hence we may write

$$I_C = I_0 + A \sin \omega t + B \sin 2\omega t + C \sin 3\omega t + D \sin 4\omega t \qquad (4.5)$$

where harmonic components up to the fourth are included. In general, harmonics of higher order than the third may be neglected. By sampling both input and output waveforms at coincident times, the co-efficients A, B, C, etc, may be determined though the work may be heavy when harmonics greater than the third are considered. The total harmonic content in a wave can be expressed as the ratio of the r.m.s. value of all harmonics to the effective value of the fundamental, that is

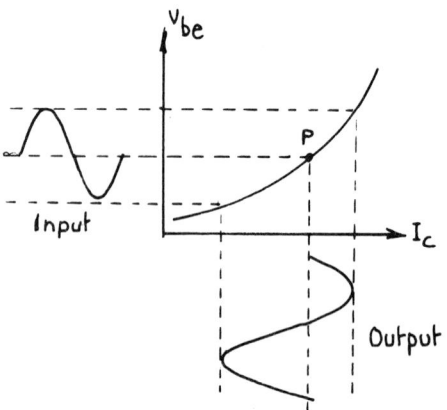

Figure 4.24

$$D_T = \frac{\sqrt{A^2 + B^2 + C^2 \cdots}}{A} \times 100\%$$

More usually the ratio of the harmonic amplitude to the fundamental amplitude is of concern. Then

% second harmonic $= \dfrac{B}{A} \times 100\%$

% third harmonic $= \dfrac{C}{A} \times 100\%$

and so on. The work is eased if the distortion can be identified as
being predominantly second or third harmonic. In general, if successive half-cycles are dissimilar as in Fig. 4.24, second harmonic
(and other even-order harmonics) are present and dominant. Such
waveforms can be obtained by adding the d.c. component I_o, the fundamental component A cos ωt and the second harmonic component B cos 2ωt.
Equation (4.5) can then be condensed to

$$I_C = I_o + A \cos \omega t + B \cos 2\omega t \tag{4.6}$$

For no signal input $I_C = I_o = I_Q$, the mean quiescent current.

Let us take samples at ωt = 0, π/2 and π with the corresponding
collector currents designated $\hat{I}$, I_o and $\check{I}$ respectively. This operation is shown in Fig. 4.25. When ωt = 0, $I_C = \hat{I}$, also cos ωt and

112

$\cos 2\omega t = 1$. Hence

$$\hat{I} = I_0 + A + B \tag{4.7}$$

When $\omega t = \pi/2$, $I_C = 0$, $\cos \omega t = 0$ and $\cos 2\omega t = -1$. Hence

$$I_Q = I_0 - B \tag{4.8}$$

When $\omega t = \pi$, $I_C = \check{I}$, $\cos \omega t = -1$ and $\cos 2\omega t = 1$. Hence

$$\check{I} = I_0 - A + B \tag{4.9}$$

Solving these last three equations simultaneously we obtain I_0, A and B in terms of $\hat{I}$, $\check{I}$ and I_Q which are known.

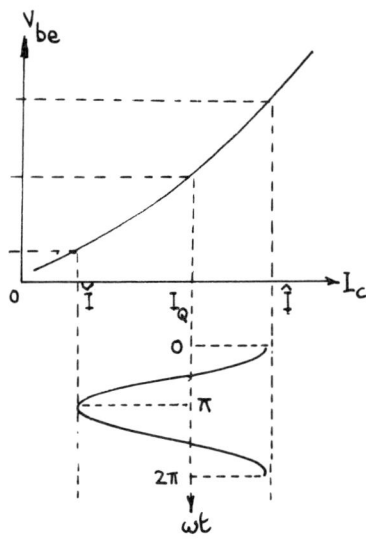

Figure 4.25

Subtracting (4.9) from (4.7) gives $(\hat{I} - \check{I}) = 2A$, so that

$$A = \tfrac{1}{2}(\hat{I} - \check{I}) \tag{4.10}$$

Adding (4.8) to (4.7) and substituting for A from (4.10) gives us

$$I_0 = \frac{\hat{I} + \check{I} + 2I_Q}{4} \tag{4.11}$$

Substituting (4.11) into (4.8) finally yields

113

$$B = \frac{\hat{I} + \check{I} + 2I_Q}{4}$$

The second harmonic distortion is then

$$\frac{B}{A} \times 100\% = \frac{\hat{I} + \check{I} - 2I_Q}{2(\hat{I} - \check{I})} \qquad (4.12)$$

Example 4.7   Obtain an expression for the shift in the d.c. output current level when a sinusoidal input voltage v = V sin ωt is applied to the amplifier input, assuming that the transfer characteristic follows a square law relationship $I_C = I_o + Av + Bv^2 + Cv^3 \ldots$

As before, with no signal input, $I_C = I_o = I_Q$. With the signal applied

$$I_C = I_Q + A\hat{V} \sin \omega t + B\hat{V}^2 \sin^2\omega t + C\hat{V}^3 \sin^3\omega t + \ldots$$

Now the mean value of sin ωt, $\sin^3\omega t \ldots$ over a cycle is zero and the mean value of $\sin^2\omega t$ is ½. Therefore, ignoring terms in the series greater than $\sin^2\omega t$

$$I_C = I_Q + \tfrac{1}{2}B\hat{V}^2$$

This is the new d.c. component; the shift in the d.c. level of the output is therefore $\tfrac{1}{2}BV^2$.

Example 4.8   A power amplifier operating with an 8Ω load has a quiescent collector current of 1.5 A. During one half of the input cycle the signal current rises to 3.1 A and during the second half-cycle it falls to 0.1 A. Calculate the power output at the fundamental frequency and the percent second harmonic distortion.

From Example 4.7 we have

$$I_C = I_Q + A\hat{V} \sin \omega t + B\hat{V}^2\sin^2\omega t$$

$$= I_Q + A\hat{V} \sin \omega t + \tfrac{1}{2}B\hat{V}^2(1 - \cos 2\omega t)$$

$$= (I_Q + \tfrac{1}{2}B\hat{V}^2) + A\hat{V} \sin \omega t - \tfrac{1}{2}B\hat{V}^2\cos 2\omega t$$

These three terms represent the d.c. component, the fundamental and the second harmonic respectively. These components are sketched in Fig. 4.26 together with their resultant output wave.

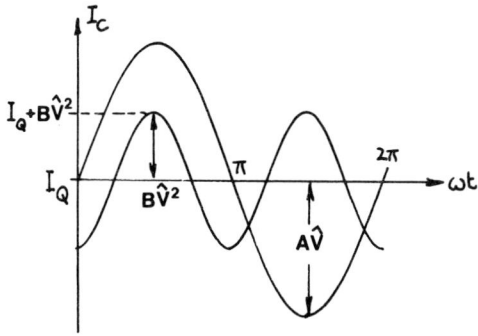

Figure 4.26

We can calculate the harmonic distortion by using equation (4.12).
Here I_Q = 1.5A, $\hat{I}$ = 3.1 A and $\check{I}$ = 0.1 A.

Therefore % second harmonic $= \dfrac{3.1 + 0.1 - (2 \times 1.5)}{2(3.1 - 0.1)}$ 100%

$$= \frac{0.2}{6} = 3.33\%$$

The 'fundamental' power = mean squared fundamental current × load

Peak fundamental current = A = $\frac{1}{2}(\hat{I} - \check{I})$

Mean squared fundamental current $= \dfrac{1}{2}[\dfrac{\hat{I} - \check{I}}{2}]^2$

Fundamental power $= \dfrac{1}{2}[\dfrac{\hat{I} - \check{I}}{2}]^2 R_L = \dfrac{(3.1 - 0.1)^2}{8} \times 8$

$$= 9 \text{ W}$$

4.6.1 Cross-over distortion

We have already noted that the operating points for the transistors
of a Class-B amplifier are not placed exactly at cut-off but are
placed slightly ahead of cut-off so that a small collector current
flows under no-signal conditions. The reason for this is to eliminate a form of distortion peculiar to Class-B amplifiers. A dynamic
transfer characteristic for a Class-B stage may be constructed by
combining the dynamic transfer characteristics for the individual

115

transistors as shown in Fig. 4.27. Because of the non-linearity in
the individual characteristics for low collector current, the operating point is chosen to be at that point ahead of cut-off where the
composite characteristic is substantially linear. The point at
which the composite characteristic crosses the v_{BE} axis is known as
the projected cut-off point. The effect of biasing the transistors
exactly at cut-off is illustrated in Fig. 4.27. Here any one of the
transistors is not fully turned on by the time the other is turned
off. As the figure shows, distortion takes place between the two
half-cycles of output or during the 'cross-over period'. The trouble
becomes worse as the signal amplitude decreases.

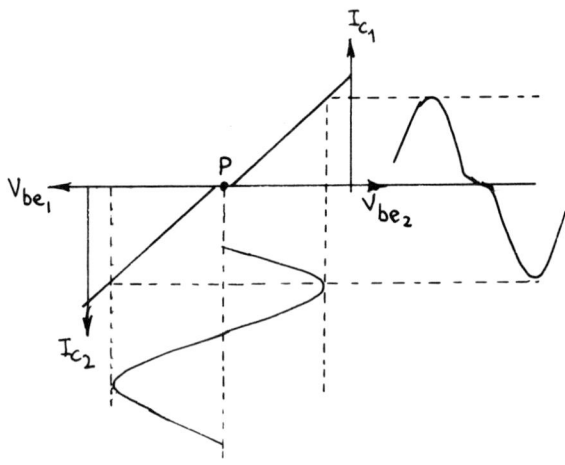

Figure 4.27

The small forward bias which effectively brings the individual
characteristics into line eliminates the cross-over distortion because as one transistor is being driven on to the non-linear region
the other is already well into conduction. This form of bias which
is ahead of cut-off is strictly called Class-AB since for 'small'
input amplitudes the operation approximates to Class-A working. The
method of course reduces the efficiency of the amplifier slightly.

If you turn back to Fig. 4.17 you will see how such a small forward bias eliminates the crossover distortion.

Circuit systems have been devised which use Class A output with
its low distortion but which avoid the large quiescent power con-
sumption by operating at a very low collector voltage 'during'

quiescent periods. This is done by using a Class-B amplifier (on
the same chassis) to provide the Class-A system with its collector
supply votage, and this voltage varies with the amplitude of the
input signal. In this way the amplifier operates in Class-A mode,
but the dissipation remains nearly the same as that of a Class-B
amplifier, as the supply voltage 'follows' the input signal amplitude.

4.7   THERMAL CONSIDERATIONS

Heat energy flows from a point at a high temperature to a point at a
lower temperature. The greater the temperature difference, the
greater the 'rate' of heat flow. As the temperature levels approach
each other, the rate decreases and the exchange is zero when the temperatures are equal. Hence the rate of heat transfer is a function
of the temperature gradient just as the rate of charge transfer (current) is a function of the potential gradient between two circuit
points. However, just as in the case of electrical conduction,
heat conduction is influenced by the 'thermal resistance' of the
medium through which it flows; good heat conductors like copper (and
most metals) have small thermal resistances, while poor heat conductors like air, wood or glass have high thermal resistances.

When a transistor dissipates internal power, heat is generated.
If the heating is very small, as it is in small-signal amplifiers,
it is usually sufficient to rely on air convection and the conducting paths provided by the connecting leads to remove the heat from
the junction as fast as it is generated. With power transistors,
however, it is not possible to conduct the heat away sufficiently
quickly by such a method and assistance has to be provided. Such
assistance takes the form of a large area of metal plate (or the
chassis may be used) which serves to dissipate the generated heat
into the surrounding environment. Such a metal plate is known as a
'heat sink'. As a further step, the collector of a power transistor
is often connected both mechanically and electrically to the transistor housing, this housing in turn being securely fastened to the
heat sink. It may be necessary to insulate (electrically) the collector from the heat sink but this can be done by the introduction
of a thin mica or silicone rubber washer.

4.7.1   Thermal resistance

The limiting factor to the build up of internal heat is the maximum permissible rated junction temperature specified for the particular transistor being used, $T_{j\ max}$. This ranges from about $80^{\circ}C - 90^{\circ}C$ for germanium and from about $140^{\circ}C - 170^{\circ}C$ for silicon. Manufacturers provide a 'derating curve' for their power transistors which shows the maximum permissible collector dissipation, $P_{C\ max}$, as a function of the ambient temperature T_a. Fig. 4.28 shows a typical curve for a small power device. From the form of this curve, the steady state temperature rise at the junction is proportional to the power dissipated, that is

$$\delta T_j = T_j - T_a \propto P_C$$

Hence

$$T_j - T_a = \theta_i P_C$$

where θ_i is a constant representing the 'intrinsic' thermal resistance of the heat path through the transistor. Strictly, θ_i is made up of two parts: the thermal resistance of the junction itself and the thermal resistance of the case to which it is attached. Hence the name 'intrinsic' as its value depends only upon the size and form of the transistor and upon the convection and radiation properties of the case to its surroundings. It is usual to refer to the intrinsic resistance as θ_{j-c}. This resistance can be measured by attaching the transistor to a metal block of sufficient size that the case can be considered to be held at the ambient temperature T_a i.e. an 'infinite' heat sink. In practical cases where the case cannot be maintained at ambient temperature and where intermediate materials such as insulating washers may be used, the total thermal resistance, θ_T, from junction to ambient has to be calculated. Clearly, the intrinsic resistance cannot be influenced by the user of the transistor.

We have already remarked that heat conduction is analogous to electrical conduction. Similarly, thermal resistance can be treated as analogous to electrical resistance; hence an 'equivalent' thermal circuit can be established from which all basic calculations may

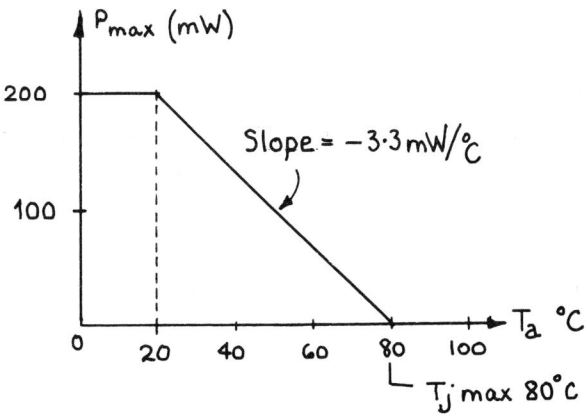

Figure 4.28

be made. Electrical resistance has the units volts-per-ampere, the
voltage dropped across a unit resistance being equal in magnitude
to the current flowing. In the same way, since

$$\text{thermal resistance } \theta = \frac{T_j - T_a}{P_C} \tag{4.13}$$

θ is defined as the temperature rise for unit power dissipation, and
from equation (4.13) is seen to have the units of degrees C per watt.

It is usual for T_j to be stated in terms of its rated maximum
value i.e. about 85°C for germanium and 150°C for silicon.

<u>Example 4.9</u>   A certain silicon transistor having $T_{j\ max}$ rated at
150°C will dissipate 25 W when its case is maintained at 100°C. What
is its intrinsic thermal resistance?

$$\theta_i = \theta_{j-c} = \frac{T_{j\ max} - T_a}{P_C}$$

$$= \frac{150 - 100}{25} = 2°C/W$$

Notice that we have taken T_a as 100°C. This is obviously not the
'actual' ambient temperature of the air surrounding the device, but
100°C is the case temperature and we are interested in the temperature 'gradient' across the transistor. The result is then the

119

required intrinsic resistance of the transistor.

The effect of the addition of a heat sink will be illustrated in the next example.

Example 4.10   A piece of blackened aluminium plate has a thermal resistance θ_s = 2.5°C/W. The transistor of Example 4.9 is bolted to it. Find the maximum collector dissipation that can be tolerated if T_a = 23°C.

The total thermal resistance from junction to ambient is the sum of the individual thermal resistances making up the conducting path, that is

$$\theta_T = \theta_{j-c} + \theta_s$$

$$= 2.0 + 2.5 = 4.5°C/W$$

The overall temperature drop is 150 - 23 = 127°C

$$P_{C \text{ max}} = \frac{127}{4.5} = 28.2 \text{ W}$$

Suppose now that a mica washer having a thermal resistance θ_w is interposed between the transistor and the heat sink, θ_w being 0.75°C/W. The total thermal resistance now becomes 2.0 + 2.5 + 0.75 = 5.25°C/W and with the same temperature gradient

$$P_{C \text{ max}} = \frac{127}{5.25} = 24.2 \text{ W}$$

Notice the reduction in the permissible dissipation resulting from the inclusion of the mica washer.

4.7.2   Thermal circuits

Problems of the kind just covered are best illustrated by applying a thermal circuit 'Ohm's law' to the representation of the heat circuit and Example 4.10 can be so shown, see Fig. 4.29. Here power dissipation due to heat flow is comparable to 'current', temperature difference is comparable to 'potential difference' and thermal resistance takes the place of electrical resistance. Notice how the intermediate temperatures are found, in exactly the same way that voltage drops are found in electrical circuits, by subtracting the

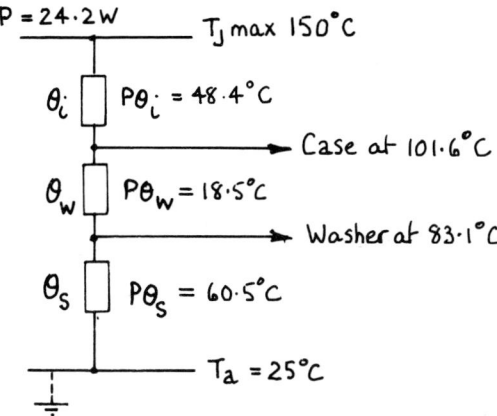

Figure 4.29

(watts times thermal resistance) 'drops' from the 'supply' temperature, $T_{j\ max} - T_a$. Notice also that, as a check, the sum of these
intermediate temperature drops comes to the supply temperature, 150°C,
within the accuracy involved.

It might be thought at first that the addition of thermal resistance in the form of a heat sink between the transistor case and the
surrounding air would lengthen the resistance path and make matters
worse rather than better. This is not so, for the thermal resistance between case and air is very large; air is a poor conductor.
The heat sink, however, effectively 'parallels' the case-to-air resistance and hence reduces its value. Figure 4.30 illustrates this
situation, θ_{c-a} being the thermal resistance from case to air. Hence
from this diagram, the total resistance

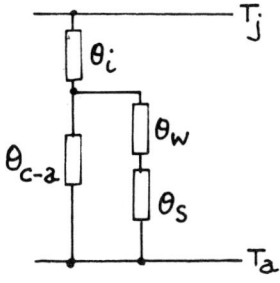

Figure 4.30

$$\theta_T = \theta_{j-c} + \frac{\theta_{c-a}(\theta_w + \theta_s)}{\theta_{c-a} + \theta_w + \theta_s}$$

where the usual rule for parallel resistances is applied.

In practical cases, a low power transistor with a rating of a few hundred milliwatts may have $\theta_i = \theta_{j-c} \simeq \theta_{c-a}$; a flag-type clip-on heat sink has θ_s about 50°C/W and when this is in a parallel with $\theta_{c-a} \simeq 200$°C.W makes the total resistance about 250°C/W instead of 400°C/W, thus almost doubling the maximum dissipation for the device. Large power transistors may have θ_{j-c} about 1°C/W and θ_{c-a} about 50°C/W, so that a large increase in rating is obtained by reducing θ_{c-a} by a parallel heat sink of resistance θ_s. It is not usually difficult to reduce θ_{c-a} by as much as nine-tenths, yielding a ten-fold increase in power rating.

Example 4.11   A silicon power transistor operating in Class-A works in an ambient temperature of 24°C. In this circumstance it is to have a maximum junction temperature of 120°C when its collector dissipation is 15 W. What is the least area of heat sink required when the material of which it is made dissipates 1.6×10^{-3} W per cm^2 for each °C above 24°C? Take θ_{j-c} to be 0.75°C/W.

No insulating washer is mentioned in this problem, so the thermal diagram may be drawn as shown in Fig. 4.31.   Then

$$T_j - T_a = \theta_T P_C$$

$$\theta_T = \frac{120 - 24}{15} = 6.4°C/W$$

$$\theta_s = \theta_T - \theta_{j-c} = 6.4 - 0.75 = 5.65°C/W$$

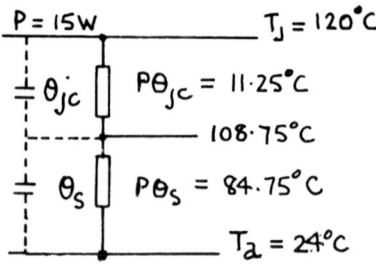

Figure 4.31

Also

$$\theta_{j-c} P_C = 15 \times 0.75 = 11.25°C$$

$$T_s = 120 - 11.25 = 108.75°C$$

Hence the sink is $108.75 - 24 = 84.75°C$ above T_a.

Let the area of the heat sink be A cm^2, then the power dissipated by the sink will be

$$P = 1.6 \times 10^{-3} \times 84.75 \times A = 15$$

Therefore

$$A = 110.6 \text{ cm}^2$$

This area would include BOTH sides of the plate.

Practical heat sinks are generally finned to provide the required surface area in a smaller bulk; they are also blackened to assist in their radiation properties.

4.7.3   Thermal time constant

When a resistive circuit is switched off the current falls to zero immediately. If the circuit contains capacitance or inductance, the current does not fall to zero until some finite time has elapsed which depends upon the circuit time-constant. In thermal circuits the time-constant can be very large. Strictly to illustrate such thermal time lag, capacitors are shown in shunt with each of the thermal resistances as Fig. 4.31 illustrates. The thermal capacitance of each part of the circuit will be very different from the others; the mass of the collector junction itself will be small so that its heat capacity and hence its thermal capacity will be small. Its time constant may therefore be only a small fraction of a second. The heat sink on the other hand will have a large mass, hence its thermal capacity will be large also and its time-constant long, probably several minutes. The importance of thermal time-constant is that the load line may pass through the area above the maximum dissipation hyperbola providing the 'excessive' dissipation does not exceed the time constant of the collector junction. Class-B amplifiers may well work in such a situation. Therefore a transistor which is specified to work with a heat sink will operate for a short while without one, but disaster is very close at hand. In the same way, the large thermal capacity of an inadequate heat sink will allow the transistor to operate for perhaps several minutes before damage results.

It does not follow from all this that a transistor is completely safe even with an adequate heat sink. Thermal runaway can result from the build up in collector current caused by an increase in I_{CBO} in the manner illustrated in Fig. 4.32. This is a feedback system; assume that the collector junction temperature T_j increases, leakage current I_{CBO} then increases and hence I_C increases. This last increase in turn raises T_j, hence thermal runaway occurs when the rate of increase of T_j exceeds the ability of the heat sink to remove the heat. Consider the small changes marked on the diagram; we have already encountered two of the connecting links θ_T and S. θ_T is the total thermal resistance between junction and ambient and S is the stability factor. The remaining links have their appropriate ratios marked on the figure and are not amenable to outside adjustment or modification. By breaking the feedback loop at any point, it is clear that the thermal loop gain is the product $aSb\theta$. For stability this must be less than unity.

Therefore

$$aSb < \frac{1}{\theta}$$

or

$$\frac{\delta P_C}{\delta T_j} < \frac{1}{\theta} \qquad (4.14)$$

This is a criterion for the avoidance of thermal runaway. By considering the diagram we can obtain an alternative form for this criterion.

The power dissipated at the junction is $P_C = I_C V_{CE}$. For a small

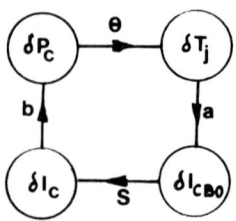

Figure 4.32

change δI_C in collector current, V_{CE} can be assumed constant (as it is in transformer coupled or complementary-symmetry amplifiers).

$$\frac{\delta P_c}{\delta I_c} = V_{CE} \text{ or } b = V_{CE} \simeq V_{CC}$$

Also

$$I_{CBO} = I_o \cdot \exp k(\delta T_j)$$

where I_o is the leakage current at T_a. Taking I_{CBO} to double for every 10^oC increase in T_j, then

$$2I_{CBO} = I_o \cdot \exp (10k)$$

$$k = 0.1 \ln 2 = 0.07$$

$$a = \frac{\delta I_{CBO}}{\delta T_j} = 0.07 \ I_{CBO}$$

Hence for stability $S.V_{CC}(0.07 \ I_{CBO}) < \frac{1}{\theta}$

or $\quad S < \dfrac{14.28}{\theta_T V_{CC} I_{CBO}}$ ⠀⠀⠀⠀⠀⠀⠀⠀⠀⠀⠀⠀⠀⠀⠀⠀⠀(4.15)

Example 4.12   Two transistors are used in a Class-B push-pull amplifier and are required to deliver a continuous output power of 30 W
and operate from a 12 V supply. If the collector voltage swing is
sinusoidal with a peak value of 10.5 V, determine the maximum permissible thermal resistance of the heat sinks used. Ratings are:
$\theta_i = 3.0^oC/W$, $\theta_{c-a} = 10^oC/W$, $T_{j \ max} = 85^oC$, $T_{a \ max} = 30^oC$.
    The total power output = 30 W = 15 W per device; for a collector
swing of 10.5 V we may take the minimum collector voltage as 1.5 V.
Then

$$\eta = \frac{\pi}{4}[1 - \frac{1.5}{12}] = \frac{7\pi}{32} = 0.69$$

$$P_c = P_L[\frac{1}{\eta} - 1] = 15[\frac{32}{7\pi} - 1]$$

$$= 6.83 \text{ W}$$

In thermal equilibrium the total thermal resistance $\theta_T = (85 - 30)/6.83 = 8.05^\circ C/W$. But from the thermal diagram of Fig. 4.33 we have

$$\theta_T = \theta_i + \frac{\theta_{c-a} . \theta_s}{\theta_{c-a} + \theta_s}$$

Therefore

$$8.05 = 3.0 + \frac{10\theta_s}{10 + \theta_s}$$

and from this we find $\theta_s = 10.2^\circ C/W$.

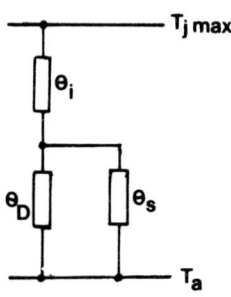

Figure 4.33

PROBLEMS 4

1. What do you understand by the term 'large signal'?

2. Why are small-signal parameters not suitable for the analysis of power amplifiers?

3. Why is impedance matching particularly necessary in power amplifiers?

4. Sketch a set of collector characteristics for a power transistor (bipolar or FET); on your diagram indicate those regions of high amplitude distortion.

5. Distinguish between the d.c. and the a.c load line for a transformer-coupled power amplifier.

6. An output transformer having a ratio of 5 to 1 step-down connects

an output power transistor to a 3 Ω load. What impedance is seen at the transistor collector?

7. Explain why a Class-A power transistor runs cooler as the power output increases.

8. Explain how push-pull amplifiers operate. Why is Class-B operation more efficient than Class-A?

9. Prove that the maximum theoretical efficiency of a Class-B amplifier is 78.5%.

10. Explain thermal runaway, illustrating your answer with a feedback diagram.

11. The maximum power hyperbola and the operating load line for a power amplifier stage are shown in Fig. 4.34. Explain why this amplifier will possibly operate satisfactorily when driven but will probably quickly fail if the drive is removed.

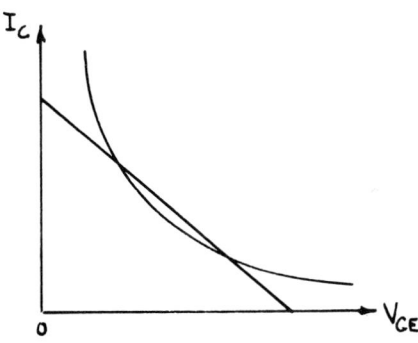

Figure 4.34

12. A Class-A power amplifier draws a quiescent collector current of 550 mA from a 20 V supply and delivers a signal power of 3.8 W to the collector load. Calculate the stage efficiency and the collector dissipation.

13. Fifteen watts of audio power is to be supplied to a 100 Ω resistive load. What output voltage and current (r.m.s) are required? For this voltage excursion, the power amplifier is driven into the

extremes of its characteristics and distortion occurs in the form of a 2.5 V second harmonic. What is this in terms of a percentage distortion? If the voltage gain of the power amplifier is 40 and the input signal to the pre-amplifier is 2 mV, what should be the gain of the pre-amplifier and what minimum number of stages might be involved?

14. A power amplifier operating in Class-A with a load of 10Ω referred to the primary of the output transformer has a quiescent collector current of 1.7 A. During one-half of the input sinusoidal cycle it is driven to $I_{C(max)}$ = 3.1 A where it saturates with $V_{CE(sat)}$ = 0.5 V. During the alternate half-cycle $I_{C(min)}$ = 0.1 A. Obtain

    (a)   the fundamental power output

    (b)   the percentage second harmonic distortion

    (c)   the supply voltage (V_{CC})

    (d)   the maximum instantaneous collector dissipation

    (e)   The maximum permissible thermal resistance between the junction and the surroundings if the junction temperature is not to exceed 140°C in an ambient temperature of 25°C.

15. Define thermal resistance as applied to the mounting of transistors, illustrating your answer in the form of a thermal-electrical analogy.

    A silicon output transistor with $T_{j(max)}$ = 150°C will dissipate 40 W when its case is maintained at 100°C. It is to be mounted on a blackened aluminium heatsink having a thermal resistance of 2.55°C/W and insulated from it by a mica washer of thermal resistance 1.2°C/W. Find the maximum power the transistor will dissipate if the ambient temperature is 25°C. Find also the case and under-washer temperatures and illustrate your answer with a suitable diagram.

16. Two transistors are operated in Class-B push-pull as power amplifiers. The supply voltage is 20 V and the circuit is required to deliver a continuous sinusoidal power output of 24 W. If the output voltage swing is 36 V peak-to-peak, calculate the maximum permissible thermal resistance of the heatsink for each device if the ambient temperature will not exceed 30°C. The mounting hardware has a

thermal resistance of 1.5°C/W and the transistor ratings are
θ_{j-c} = 3.5°C/W, $T_{j(max)}$ = 80°C.

17. Assuming that the rate of removal of heat from a junction is proportional to the temperature difference between junction and ambient,
derive an expression for the thermal resistance from junction to air
as a function of the collector dissipation and the temperature difference between junction and ambient.

18. The current flowing in the centre-tap supply lead to the transformer of a Class-B power amplifier is 750 mA. If the transformer
ration is 1 : 1 and the load is 40 Ω, what power is being delivered
to the load? Assume that circuit losses are negligible.

19. The derating curve of a power transistor is given in Fig. 4.35.
Discuss the meaning and relevance of this curve.

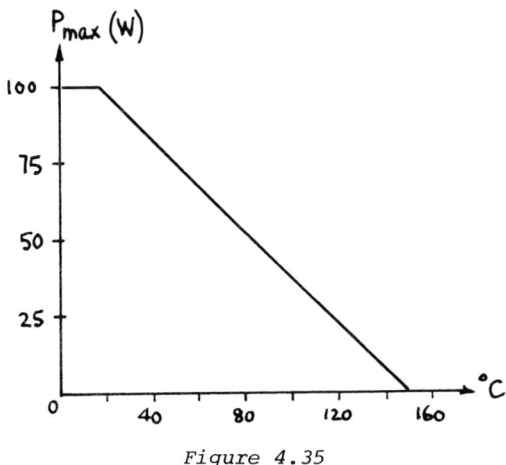

Figure 4.35

The transistor concerned is mounted on a heat sink the material
of which has a thermal resistance of 30°C/W per 10 cm^2 of surface
area, and is insulated from it by a mica washer of thermal resistance
1.3°C/W. If the junction temperature is to be limited to 120°C obtain
(i) the intrinsic thermal resistance of the transistor, (ii) the maximum power dissipation, (iii) the least surface area of heat sink required, given T_a = 20°C.

20. The output characteristics of a power transistor are provided in
Fig. 4.36. This transistor is operated as a Class-A amplifier,

coupled to a load resistor of 0.4Ω through an ideal transformer of turns ratio 4.9 : 1 and with V_{CC} = 12.5 V. Draw on the characteristics (in pencil) the d.c. load line and the a.c. load line under the stated conditions.

An input signal of peak-to-peak amplitude 50 mA is applied to the base of the transistor. If the signal is sinusoidal and the input resistance of the transistor is 10 Ω, estimate (a) the power supplied to the load, (b) the power gain in dB, (c) the collector efficiency.

The maximum collector dissipation for this transistor is stated to be 16 W. By tracing the maximum power hyperbola on the given characteristics, comment on the suitability, or otherwise, of the chosen operating point.

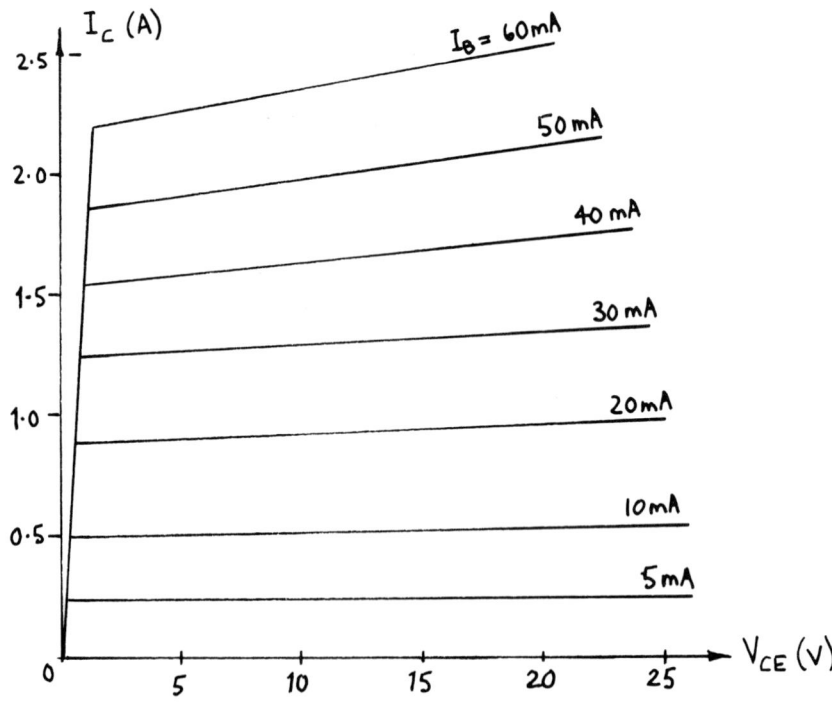

Figure 4.36

21. Define the terms (a) derating factor, (b) thermal resistance, (c) stability factor. A given transistor has the following ratings: $T_{j\ max}$ = 95°C, $P_{c\ max}$ = 150 mW at 20°C, I_{cbo} = 2.5 μA at 20°C.

Evaluate for this device (i) the derating factor, (ii) the thermal resistance.

The collector characteristics for the above transistor are given in Fig. 4.37. This transistor is to operate in common-emitter mode along the load line shown. If the operating point is to remain within the limits indicated by the points P_1 and P_2 as the ambient temperature changes from 20°C to 50°C respectively, obtain a figure for the stability factor required for this circuit, assuming that I_{cbo} doubles for each 10°C rise in junction temperature.

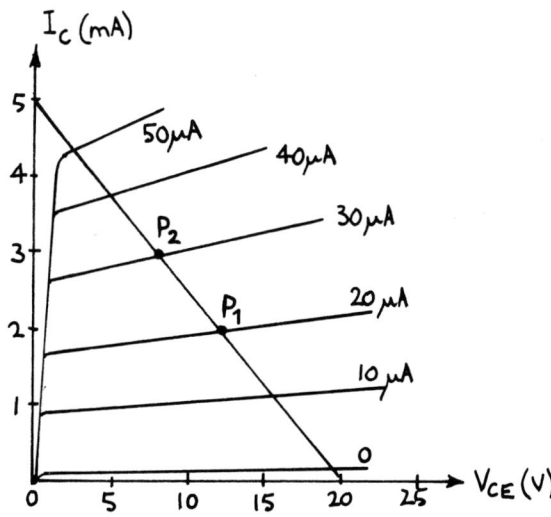

Figure 4.37

5 Small signal models

5.1 INTRODUCTION

By definition, a 'small signal' is one to which an electronic device
responds in a linear manner. In small-signal amplifiers, therefore,
we may limit our considerations to small signals superimposed on d.c.
values and assume linear relationships among signals in devices that
are decidedly non-linear in their large-signal behaviour. On this
basis, small-signal amplifiers are best dealt with by the use of circuit models or equivalents, so making possible an analysis in terms
of systems composed of ideal linear elements. While some of these
equivalent circuits are based on the physical processes going on
within a transistor, others do not necessarily represent in any way
such actual physical interpretation, but they do nevertheless represent linear 'models' which are amenable to mathematical analysis.
From such models, accurate predictions of the behaviour of the devices
in practical applications máy then be made without the labours involved in graphical analysis or the problems posed by the wide variations experienced in the characteristic curves of individual devices
of similar type.

5.2 NETWORK ANALYSIS

In the elementary amplifier of Fig. 5.1(a) the input signal v_1 is
superimposed on the base-emitter bias voltage V_{BB}. The signal current
i_1 controls the flow of collector current i_2 and this current develops an output voltage v_2 across the load resistor R_L. For purposes of small-signal analysis the actual circuit within the broken
line of Fig. 5.1(a) may be replaced by the two-port (or four-terminal)

132

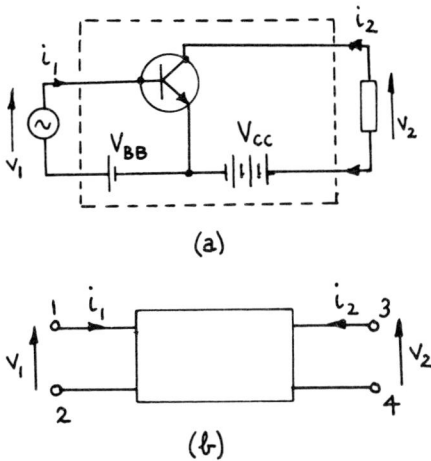

(a)

(b)

Figure 5.1

box shown in Fig. 5.1(b), where the box is considered to contain a simple equivalent circuit to that of the actual amplifier. The equivalent circuit is concerned only with a.c. (signal) conditions; all d.c. sources are considered to have zero a.c. impedance and are replaced by a.c. short-circuits.

The conventions used for the input and output currents and voltages are those shown, the sense of the currents and voltages being chosen so that the system is symmetrical and neither terminals 1, 2 nor terminals 3, 4 are specifically input or output. There are six possible ways of setting up equations to relate the four quantities and the parameter description of the network depends upon which pair of these four quantities are selected as the independent variables. Of the six possibilities only three will be of direct interest and only one of these will be developed in detail.

5.2.1   Impedance parameters

The impedance (or z-parameters) equations of a network may be defined by taking current i_1 and i_2 as the independent variables. The behaviour of the circuit can then be defined by the functional relationships

$$v_1 = f(i_1, i_2)$$

133

$$v_2 = f(i_1 \cdot i_2)$$

Consider differential changes in the variables, since we are dealing with small-signal amplitudes:

$$dv_1 = \frac{\partial v_1}{\partial i_1} di_1 + \frac{\partial v_1}{\partial i_2} di_2 \qquad\qquad (5.1(\text{i}))$$

$$dv_2 = \frac{\partial v_2}{\partial i_1} di_1 + \frac{\partial v_2}{\partial i_2} di_2 \qquad\qquad (5.1(\text{ii}))$$

The first of these tells us that a change in input current di_1 and a change in output current di_2 both contribute to a change in input voltage dv_1. What the contributions are depends upon the co-efficients $\partial v_1/\partial i_1$ and $\partial v_1/\partial i_2$. These partial derivatives of input voltage with respect to input current and output current, and the corresponding partial derivatives of output voltage in the second equation are the z-parameters of the small-signal equivalent circuit.

Noting that the differential changes dv_1 and di_2 correspond to 'small-signal' quantities v_1 and i_2, we may write the equations of (5.1) in the form

$$v_1 = \frac{\partial v_1}{\partial i_1} i_1 + \frac{\partial v_1}{\partial i_2} i_2 = z_i i_1 + z_r i_2 \qquad\qquad (5.2(\text{i}))$$

$$v_2 = \frac{\partial v_2}{\partial i_1} i_1 + \frac{\partial v_2}{\partial i_2} i_2 = z_f i_1 + z_o i_2 \qquad\qquad (5.2(\text{ii}))$$

where z_i, z_r, z_f and z_o are impedances. The terms $z_i i_1$ and $z_o i_2$ are voltage drops while the terms $z_r i_2$ and $z_f i_1$ require the use of de-pendent voltage generators whose e.m.f.s are proportional to the out-put and input currents respectively. Since the voltage terms are added, the input and output circuits are of series form. From this line of reasoning the z-parameter model of Fig. 5.2 follows. The in-dividual parameters may now be defined by taking the extreme cases of either $i_1 = 0$ or $i_2 = 0$. These cases correspond to the input or out-put terminals being open-circuited respectively. Hence, from equa-tion (i) in (5.2) above setting $i_2 = 0$

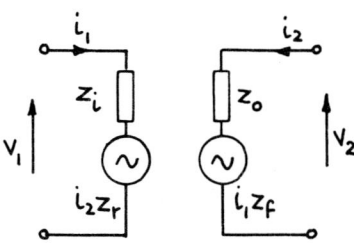

Figure 5.2

$$z_i = \frac{v_1}{i_1}\bigg|_{i_2 = 0} = \text{input impedance}$$

$$z_f = \frac{v_2}{i_1}\bigg|_{i_2 = 0} = \text{forward transfer impedance}$$

In the same way, setting $i_1 = 0$

$$z_r = \frac{v_1}{i_1}\bigg|_{i_1 = 0} = \text{reverse transfer impedance}$$

$$z_o = \frac{v_2}{i_2}\bigg|_{i_1 = 0} = \text{output impedance}$$

In matrix form we have

$$\begin{pmatrix} v_1 \\ v_2 \end{pmatrix} = \begin{pmatrix} z_i & z_r \\ z_f & z_o \end{pmatrix} \begin{pmatrix} i_1 \\ i_2 \end{pmatrix}$$

where $\begin{pmatrix} z_i & z_r \\ z_f & z_o \end{pmatrix}$ is the transfer matrix.

5.2.2   Admittance parameters

If v_1 and v_2 are chosen as independent variables, the resulting parameters are admittance or y-parameters. The defining equations are then

$$i_1 = f(v_1, v_2)$$

$$i_2 = f(v_1, v_2)$$

and

$$i_1 = \frac{\partial i_1}{\partial v_1} dv_1 + \frac{\partial i_1}{\partial v_2} dv_2 = y_i v_1 + y_r v_2$$

$$i_2 = \frac{\partial i_2}{\partial v_1} dv_1 + \frac{\partial i_2}{\partial v_2} dv_2 = y_f v_1 + y_o v_2 \qquad (5.3)$$

where

$$y_i = \left. \frac{i_1}{v_1} \right|_{v_2 = 0} = \text{input admittance}$$

$$y_f = \left. \frac{i_2}{v_1} \right|_{v_2 = 0} = \text{forward transfer admittance}$$

$$y_r = \left. \frac{i_1}{v_2} \right|_{v_1 = 0} = \text{reverse transfer admittance}$$

$$y_o = \left. \frac{i_2}{v_2} \right|_{v_1 = 0} = \text{output admittance}$$

Since each term in the defining equations (5.3) represents current, parallel combinations of a controlled current source and an admittance are indicated. The equivalent circuit may then be deduced and is shown in Fig. 5.3.

In matrix form we have

$$\begin{pmatrix} i_1 \\ i_2 \end{pmatrix} = \begin{pmatrix} y_i & y_r \\ y_f & y_o \end{pmatrix} \begin{pmatrix} v_1 \\ v_2 \end{pmatrix}$$

and the transfer matrix is

$$\begin{pmatrix} y_i & y_r \\ y_f & y_o \end{pmatrix}$$

Example 5.1   Determine the z-parameters for the network shown in

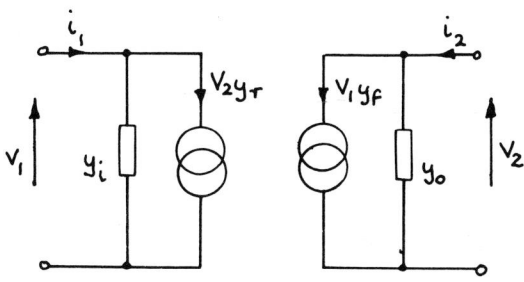

Figure 5.3

Fig. 5.4.

This is a passive network, containing no active or non-linear devices. Marking in the circulating currents i_1, i_2 and i_3 we have

$$v_1 = 4i_1 - i_3 \tag{i}$$

$$0 = -i_1 + 6i_3 + 3i_2 \tag{ii}$$

$$v_2 = 3i_3 + 8i_2 \tag{iii}$$

From (ii)

$$i_3 = \frac{1}{6}i_1 - \frac{1}{2}i_2$$

and substituting this into (i) and (iii) gives us

$$v_1 = 4i_1 - \frac{1}{6}i_1 - \frac{1}{2}i_2$$

$$= \frac{23}{24}i_1 + \frac{1}{2}i_2$$

$$v_2 = \frac{1}{2}i_i - \frac{3}{2}i_2 + 8i_2$$

$$= \frac{1}{2}i_1 + \frac{13}{2}i_2$$

Comparing these results with the defining equations for the z-parameters where

$$v_1 = z_i i_1 + z_r i_2$$

$$v_2 = z_f i_1 + z_o i_2$$

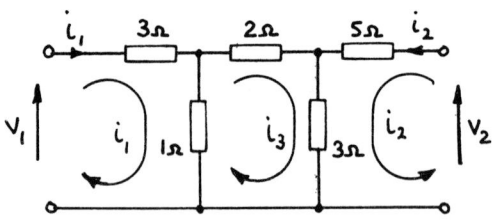

Figure 5.4

we have

$$z_i = \frac{23}{24}\ \Omega, \quad z_r = \frac{1}{2}\ \Omega, \quad z_f = \frac{1}{2}\ \Omega, \quad z_o = \frac{13}{2}\ \Omega$$

Notice that $z_f = z_r$. This will always be so for a passive network and may be used as a test for passivity in certain cases.

5.2.3 h-parameters

An equivalent circuit is chosen for the representation of transistor action in which the input circuit is of series form and the output circuit is of parallel form. This means taking v_2 and i_i as the independent variables. The chief advantage of this circuit is that it can be applied to all three transistor circuit configurations, the parameters are most frequently quoted by the manufacturers and the parameters themselves can be measured with little difficulty.

It might seem that the defining equations this time would simply consist of equation (i) from (5.2) and equation (ii) from (5.3), but these equations are interrelated. To obtain the required independent equations i_1 is made the current variable in the right-hand side of both equations and v_2 the voltage variable. Working as before, the defining equations are then

$$v_1 = h_i i_1 + h_r v_2 \qquad\qquad\qquad (5.4(i))$$

$$i_2 = h_f i_1 + h_o v_2 \qquad\qquad\qquad (5.4(ii))$$

The dimensions of the parameters are now mixed or 'hybrid' since h_i is an impedance, h_r is a voltage ratio, h_f is a current ratio and h_o is an admittance. Hence the symbol 'h' and the name hybrid parameters. So

$$h_i = \left. \frac{v_1}{i_1} \right|_{v_2 = 0} = \text{input impedance } (\Omega)$$

$$h_r = \left. \frac{v_1}{v_2} \right|_{i_1 = 0} = \text{reverse voltage transfer ratio}$$

$$h_f = \left. \frac{i_2}{i_1} \right|_{v_2 = 0} = \text{forward current transfer ratio}$$

$$h_o = \left. \frac{i_2}{v_2} \right|_{i_1 = 0} = \text{output admittance (S)}$$

From equation (i) of (5.4) the input voltage v_i is the sum of an impedance drop $h_i i_1$ and a controlled voltage generator $h_r v_2$, derived from and directly proportional to the output voltage. Equation (ii) of (5.4) indicates a parallel arrangement of an admittance current $h_o v_2$ and a controlled current generator source $h_f i_1$. The h-parameter equivalent circuit of Fig. 5.5 then follows.

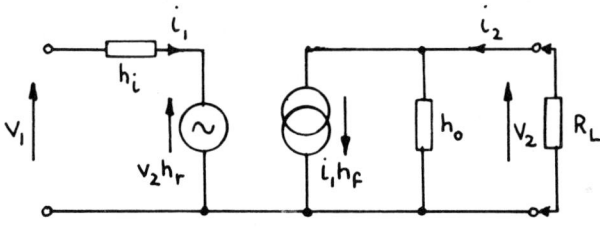

Figure 5.5

As already mentioned, this model can be used for any bipolar transistor configuration. In a particular case, subscripts are added to the parameters for purposes of mode identification, these being e, b or c depending upon whether the emitter, base or collector respectively is the common electrode. It should be added that we have ignored the internal transistor capacitances in this analysis, that is, the equivalent circuit is a low-frequency model. For high frequency work, some modification is needed and this will be discussed in due course.

The h, z or y-parameters are interrelated and the next example illustrates a case.

<u>Example 5.2</u>   Relate the h-parameters to the z-parameters.

We have

$$z_i = \frac{v_1}{i_1}\bigg|_{i_2 = 0}$$

Then eliminating v_2 from the defining equations for the h-parameters

$$v_1 = h_i i_1 + h_r v_2 \tag{i}$$

$$0 = h_f i_1 + h_o v_2 \tag{ii}$$

and

$$z_i = \frac{h_i h_o - h_r h_f}{h_o} = \frac{\Delta}{h_o}$$

We shall see that it is convenient to use Δ in place of $h_i h_o - h_r h_f$ as we proceed.

Now

$$z_f = \frac{v_2}{i_1}\bigg|_{i_2 = 0}$$

Hence from (ii) above

$$z_f = -\frac{h_f}{h_o}$$

Also

$$z_r = \frac{v_1}{i_2}\bigg|_{i_1 = 0}$$

Hence, setting $i_1 = 0$ in the defining equations, $v_1 = v_2 h_r$ and $i_2 = h_o v_2$. Hence

$$z_r = \frac{h_r}{h_o}$$

Finally

$$z_o = \frac{v_2}{i_2}\bigg|_{i_1 = 0}$$

Hence, again setting $i_1 = 0$ in the defining equations we obtain

$$z_o = \frac{1}{h_o}$$

5.3 GAIN AND IMPEDANCE ANALYSIS

We can derive general expressions for small-signal voltage and current gain together with input and output impedance by considering the h-parameter circuit model of Fig. 5.5. when a load resistor is connected across the output terminals. Restating the defining equations together with an expression for i_2 in terms of R_L we have

$$v_1 = h_i i_1 + h_r v_2 \qquad \text{(i)}$$

$$i_2 = h_f i_1 + h_o v_2 \qquad \text{(ii)}$$

$$v_2 = -R_L i_2 \qquad \text{(iii)}$$

From equations (ii) and (iii) we have

$$h_f i_1 + h_o v_2 + \frac{v_2}{R_L} = 0 \qquad \text{(iv)}$$

and rearranging this in terms of i_1 gives us

$$i_1 = -\frac{1}{h_f}[\frac{1}{R_1} + h_o]v_2$$

Substituting back into (i)

$$v_i = -\frac{h_i}{h_f}[\frac{1}{R_L} + h_o]v_2 + h_r v_2$$

Therefore

$$\frac{v_i}{v_2} = -\frac{h_i}{h_f}[\frac{1}{R_L} + h_o] + h_r$$

Then voltage gain $A_v = \frac{v_2}{v_1} = -\frac{h_f R_L}{h_i + \Delta R_L} \qquad \text{(5.5)}$

To derive current gain, we eliminate i_2 between (iii) and (iv)

$$- \frac{v_2}{R_L} = h_f i_1 + h_o v_2$$

$$v_2 = -i_1 \frac{h_f R_L}{h_o R_L + 1}$$

Therefore

$$A_i = \frac{i_2}{i_1} = \frac{h_f}{h_o R_L + 1} \tag{5.6}$$

To derive the input resistance we consider equation (i)

$$v_1 = h_i i_1 + h_r v_2$$

$$= h_i i_1 - h_r i_2 R_L$$

$$= h_i i_1 - \frac{i_1 h_f R_L h_r}{h_o R_L + 1}$$

by substituting for i_2 from equation (5.6)

Then

$$R_i = \frac{v_1}{i_1} = h_i - \frac{h_f h_r R_L}{h_o R_L + 1}$$

$$= \frac{\Delta R_L + h_i}{h_o R_L + 1} \tag{5.7}$$

Notice that the input resistance is influenced by the output load R_L.

For the output resistance, consider Fig. 5.6 where the input and output terminals are reversed in their roles. Then $R_o = v_2/i_2$. R_L has no effect here since we are looking back into the transistor terminals. Since R_i depends upon R_L, however, it is reasonable to suppose that R_o will depend upon the source resistance R_s. Hence

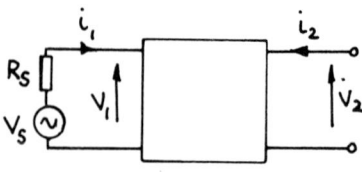

Figure 5.6

142

for a voltage v_2 applied at the output terminals and for $v_s = 0$

$$v_1 = -i_1 R_s$$

and substituting this into defining equation (i) gives

$$-i_1 R_s = h_i i_1 + h_r v_2$$

or

$$i_1 = -\frac{v_2 h_r}{h_i + R_s}$$

Substituting this into equation (ii)

$$i_2 = -v_2 \frac{h_r h_f}{h_i + R_s} + h_r v_2$$

Therefore

$$\frac{1}{R_o} = \frac{i_2}{v_2} = h_o - \frac{h_r h_f}{h_i + R_s}$$

from which

$$R_o = \frac{h_i + R_s}{\Delta + h_o R_s} \qquad (5.8)$$

Notice that the input source resistance R_s, as expected, has influenced R_o.

There is little object in trying to memorise these expressions for gain and resistance as such; each problem should be worked from the equivalent circuit model and the defining equations.

Example 5.3   The small-signal voltages and currents for a common-emitter amplifier are

$$v_1 = 2000 i_1 + 10^{-5} v_2$$

$$i_1 = 80 i_1 + 10^{-4} v_2$$

The amplifier has a collector load of 5 kΩ and is fed from a source of e.m.f. 1 mV and internal resistance 500 Ω. Draw an equivalent circuit, identify the h-parameters, and calculate the output signal voltage across R_L and the base and collector signal currents.

The equivalent circuit is shown in Fig. 5.7 where the relevant parameters are $h_{ie} = 2 \times 10^3$ Ω, $h_{re} = 10^{-5}$, $h_{fe} = 80$ and $h_{oe} = 10^{-4}$S.

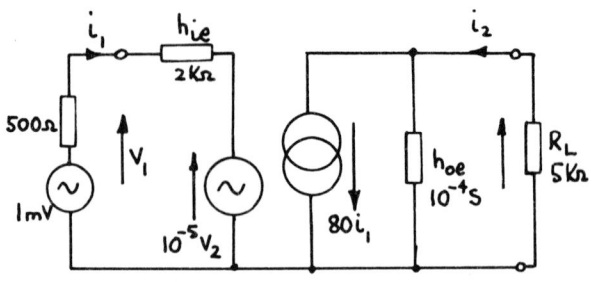

Figure 5.7

Clearly

$$v_1 = 0.8 \times 10^{-3} \text{ V}$$

Then

$$A_v = - \frac{h_{fe} R_L}{\Delta R_L + h_{ie}}$$

where $\Delta = h_i h_o - h_r h_f = (2 \times 10^3 \times 10^{-4})(80 \times 10^{-5})$

$$\simeq 0.2$$

Therefore

$$A_v = \frac{80 \times 5 \times 10^3}{(0.2 \times 5 \times 10^3) + 2 \times 10^3} = -133$$

Then

$$v_2 = A_v v_1 = -133 \times 0.8 \times 10^{-3} \text{ V}$$

$$-106 \text{ mV}$$

Also

$$i_2 = - \frac{v_2}{R_L} = \frac{106 \times 10^{-3}}{5 \times 10^3} \simeq 21 \text{ μA}$$

Now

144

$$80i_1 = i_2 - 10^{-4}v_2$$

$$= (21 \times 10^{-6}) - (-106 \times 10^{-3}) \text{ A}$$

$$= 31.6 \ \mu\text{A}$$

Hence

$$i_1 = 0.395 \ \mu\text{A}.$$

Example 5.4   The h-parameters of a transistor are h_{ie} = 1 kΩ, h_{fe} = 50, h_{re} = 5 $\times$ 10^{-4}, h_{oe} = 80 μS. Calculate the output voltage and the output resistance of a common-emitter stage using this transistor with a load R_L = 5 kΩ and fed from a source of e.m.f. 10 mV and internal resistance 300 Ω.

Fig. 5.8 shows the equivalent circuit. Working from this circuit, for the input loop

$$10^{-2} = 1300i_i + (5 \times 10^{-4})v_2 \tag{i}$$

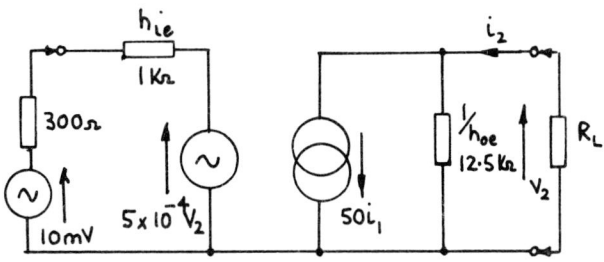

Figure 5.8

The effective load resistance is

$$R'_L = \frac{12.5 \times 5}{12.5 + 5} = 3.57 \text{ k}\Omega$$

since $1/h_o$ = 12.5 kΩ.

$$v_2 = -50 \times 3.57 \times 10^3 i_1 = -178.5 \times 10^3 i_1$$

and substituting for i_1 in (i)

145

$$10^{-2} = -\frac{1300v_2}{178.5 \times 10^3} + (5 \times 10^{-4})v_2$$

Therefore

$$v_2 = -\frac{10^{-2}}{67.5 \times 10^{-4}}$$

$$= -1.48 \text{ V}$$

To obtain the output resistance we replace the 10 mV source by its internal resistance and connect a voltage generator across the output terminals, see Fig. 5.9. From this equivalent circuit we

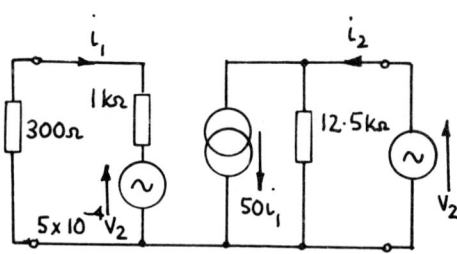

Figure 5.9

have

$$5 \times 10^{-4}v_2 = -1300i_1 \qquad \qquad \text{(ii)}$$

and

$$v_2 = (i_2 - 50i_1)(12.5 \times 10^3)$$

Substituting for i_1 from (ii)

$$v_2 = 12.5 \times 10^3i_2 + \frac{50 \times 5 \times 10^{-4} \times 12.5 \times 10^3v_2}{1300}$$

$$= 12.5 \quad 10^3i_2 + 0.24v_2$$

Therefore

$$R_o = \frac{v_2}{i_2} = \frac{12.5 \times 10^3}{0.76} \ \Omega$$

$$= 16.45 \text{ k}\Omega$$

The effective output resistance with the load connected is then

$$R_o' = \frac{16.45 \times 5}{16.45 + 5} = 3.83 \text{ k}\Omega$$

5.4 MEASUREMENT OF h-PARAMETERS

The h-parameters are small signal parameters assuming small a.c. variations about the operating points defined by d.c. values of voltage and current. The parameters could be found very approximately, therefore, by drawing tangents at suitable points on the transistor static characteristics. This is not a suitable method and in actual practise the parameters are determined experimentally by measuring the appropriate a.c. currents and voltages that result from a.c. signals applied under the implied conditions. Manufacturers' literature usually quotes the h-parameters for a number of working points and these are measured at a representative frequency of the order of 1 kHz. To determine h_{fe}, for example, a small a.c. voltage is applied between base and emitter and the resulting a.c. currents in the base lead and the collector are measured. Under the condition that

$$h_{fe} = \frac{i_2}{i_1} \Bigg|\; v_2 = 0$$

the output must be short-circuited to a.c. and this is achieved by connecting a suitable capacitor between the collector and emitter.

5.5 PARAMETER CONVERSIONS

We have derived performance equations in terms of the general parameters h_i, h_r, h_f and h_o and used these equations to calculate current and voltage gain, input and output resistance for the common-emitter configuration by using the specific parameters h_{ie}, h_{re}, h_{fe} and h_{oe}. The equivalent circuit of Fig. 5.5 is true for all three configurations because they must all satisfy the defining equations. It is therefore possible to convert from one given set of parameters, say those applicable to common-emitter configuration, to sets applicable to either common base or common-collector mode. For example, consider the common-emitter and common-collector circuits shown in Fig. 5.10(a) and (b) respectively.

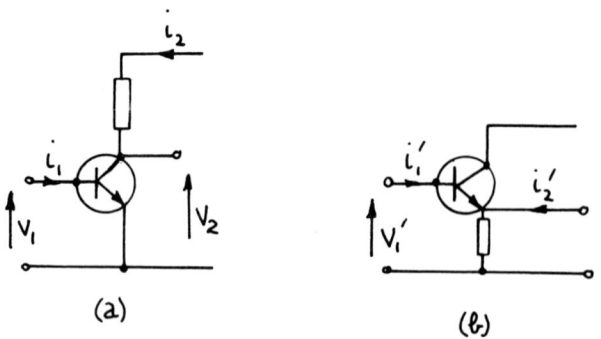

Figure 5.10

The defining equations for these modes are respectively

$$v_1 = h_{ie}i_1 + h_{re}v_2 \qquad\qquad v_1' = h_{ic}i_1' + h_{rc}v_2'$$

$$i_2 = h_{fe}i_1 + h_{oe}v_2 \qquad\qquad i_2' = h_{fe}i_1 + h_{oc}v_2'$$

Then from the circuits, since $i_e = i_c + i_b$ and $v_{eb} + v_{bc} + v_{ce} = 0$

$$i_1' = i_b = i_1$$

$$i_2' = -i_e = -(i_c + i_b) = -(i_2 + i_1)$$

$$v_2' = -v_{ce} = -v_2$$

$$v_1' = v_1 - v_2 = v_{be} + v_{ec}$$

Hence the common-collector defining equations become

$$v_1 - v_2 = h_{ic}i_1 - h_{rc}v_2$$

$$-(i_1 + i_2) = h_{fc}i_1 - h_{oc}v_2$$

Comparing these with the common-emitter equations gives

$$h_{ic} = h_{ie}$$

$$h_o = h_{oe}$$

Also

$$v_1 - v_2 = h_{re}v_2 - v_2 = (h_{re} - 1)v_2$$

Therefore

$$-h_{rc}v_2 = (h_{re} - 1)v_2$$

Therefore

$$h_{rc} = (1 - h_{re}) \simeq 1$$

Finally

$$-i_2 = -h_{fe}i_1 - h_{oe}v_2$$

Therefore

$$-i_2 - i_1 = -h_{fe}i_1 - i_1 = -i_1(h_{fe} + 1)$$

Therefore

$$h_{fc} = -(h_{fe} + 1) \simeq -h_{fe}$$

Fig. 5.11(a) and (b) shows respectively the h-parameter equivalent circuit for common-collector mode using common-collector h parameters and using common-emitter h parameters. This shows that with the output short-circuited to a.c., the common-collector

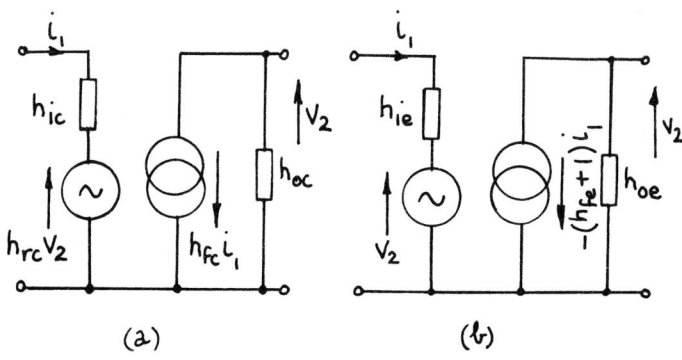

(a)   (b)

Figure 5.11

149

configuration is identical with the common-emitter configuration. As the above results have shown, the low-frequency input resistances are the same, and with the input open, the output admittances are identical. Only h_{rc} differs for the common-collector case where it is practically equal to unity in comparison with the common-emitter case where it is negligible. Notice also that the current gain for common-collector is negative. This means that if the input current flows into the transistor, the output current flows out of the transistor - a phase reversal.

In a similar manner, the relationships between common-base and common-emitter and between common-base and common-collector parameters may be established. Table 5.1 gives a summary of these last relationships; these are approximations, but are good enough for most practical applications.

TABLE 5.1   Approximate h-parameter relationships

Common-base/common-emitter	Common-base/common-collector
$h_{ie} = h_{ib}/(h_{fb} + 1)$	$h_{ic} = h_{ib}/(h_{fb} + 1)$
$h_{fe} = -h_{fb}/(h_{fb} + 1)$	$h_{fc} = -1(h_{fb} + 1)$
$h_{re} = \dfrac{h_{ib}h_{ob}}{h_{fb} + 1}$	$h_{rc} = 1$
$h_{oe} = h_{ob}(h_{fb} + 1)$	$h_{oc} = h_{ob}/(h_{fe} + 1)$

You might care to verify these results for yourself.

Example 5.5   Evaluate the common-emitter and common-collector h parameters for a transistor having the following common-base parameters: $h_{ib} = 30\ \Omega$, $h_{fb} = -0.98$, $h_{rb} = 2 \times 10^{-4}$, $h_{ob} = 0.5 \times 10^{-6}$S.
From the conversions we obtain

$$h_{ie} = \frac{30}{1 - 0.98} = 1500\ \Omega$$

$$h_{fe} = \frac{0.98}{1 - 0.98} = 49$$

$$h_{re} = \frac{30 \times 0.5 \times 10^{-6}}{1 - 0.98} - 2 \times 10^{-4}$$

$$= 0.55 \times 10^{-3}$$

$$h_{oe} = \frac{0.5 \times 10^{-6}}{1 - 0.98} = 25 \times 10^{-6} \text{ S}$$

From these results we obtain in turn

$$h_{ic} = 1500, \ h_{rc} = 1, \ h_{fc} = -50, \ h_{oc} = 25 \times 10^{-6} \text{ S}.$$

Care must be taken with the signs of the parameters when calculations are being made.

5.6   DERIVATION OF MODELS FROM PHYSICAL CONSIDERATIONS

By a consideration of the characteristics of a transistor in its various configurations, an equivalent circuit may be built up based on the physical processes within the transistor rather than on pure network theory so far pursued.   Such a model will consist of a re-sistance in each of the three branches associated with the terminals of the transistor.   To simulate the amplification provided by the transistor a current - or voltage-dependent generator is added to the circuit.

Consider first the idealised common-base characteristics shown in Fig. 5.12.   The input characteristic shown at (a) is that of a theoretical diode while the output characteristic shown at (b)

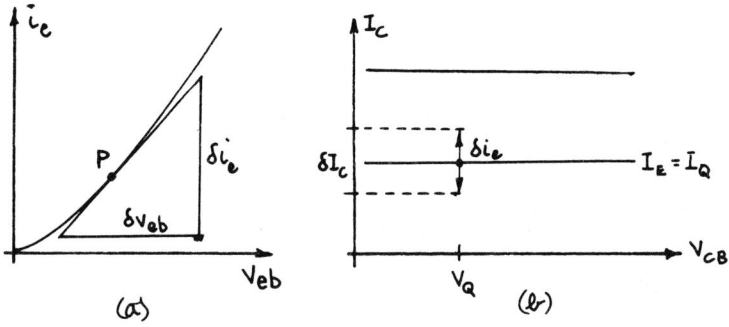

Figure 5.12

indicates a constant current gain h_{FB} ($=\alpha_B$) and negligible leakage. The gradient of the input characteristic (at a stated point) is $\delta v_{eb}/\delta i_e$ and represents the common-base a.c. input resistance which may be denoted by r_e. At room temperature, this resistance which is identical with the resistance of a forward-biased diode, can be expressed as

$$r_e = \frac{25}{I_E(\text{in mA})} \ \Omega$$

from equation (1.10) earlier.

In addition to r_e there is the incremental resistance r_b of the base which is the effective transverse ohmic resistance of the lightly doped base region. This resistance varies somewhat with collector potential because the base width varies with collector potential. As no other effect has to be accounted for, the input side of the equivalent circuit is now established.

For the output side we notice that at a quiescent point P defined by $I_E = I_Q$ and $V_{CB} = V_Q$, a small change in input voltage produces a small change δi_e in emitter current which in turn produces a small change δi_c in collector current. The ratio $\delta i_c/\delta i_e$ is the small-signal common-base current gain h_{fb}. For most practical purposes we may take $h_{fb} = h_{FB}$. As no other effect has to be accounted for, the output side of the equivalent circuit is now established. We consequently obtain the small-signal model shown in Fig. 5.13 where the output circuit contains a controlled current source $h_{fb}i_e$ shunted by the element r_c.

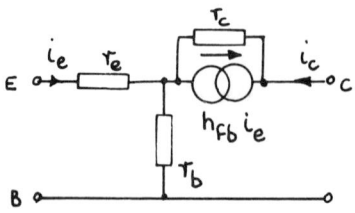

Figure 5.13

In using such an equivalent T as it is called, it is often convenient to replace the current generator by its Thevénin voltage generator equivalent. The result of this substitution (which you might
care to work out for yourself) is shown in Fig. 5.14. The value of
the source resistance r_C does not change but its position does, and
a generator of current I becomes a voltage generator of magnitude IR.

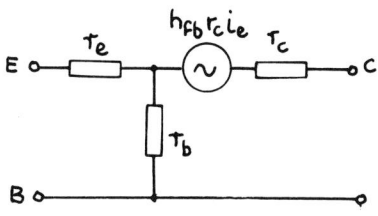

Figure 5.14

For the common-emitter connection the transistor has to be reconnected so that the base-emitter terminals become the input port
and the collector-emitter terminals the output port. This means
interchanging the branches that include r_e and r_b. This rearrangement leads to a complete change in the circuit performance.

Although the model of Fig. 5.13 could be used, it is not convenient because the controlled source is not a function of the input
current; we have to think of the collector current generator as
dependent upon the input quantity i_b rather than i_e. To change from
the current source $h_{fb}i_e$ in shunt with r_c to one of $h_{fe}i_b$ in shunt
with some new resistance r_c', we note (keeping in mind our sign
convention)

$$i_e = i_c - i_b = h_{fb}i_e - i_b$$

Then

$$i_e = \frac{-i_b}{1 - h_{fb}} \quad \text{or} \quad h_{fb}i_e = \frac{-h_{fb}}{1 - h_{fb}} i_b$$

$$= -h_{fe}i_b$$

In making the source transformation, therefore, the resulting
voltage generator can be expressed as $h_{fb}(i_b + i_c)r_c$. But in series

with this generator is r_c and consequently an $i_c r_c$ voltage drop. If we combine the rise $h_{fb} i_c r_c$ with this drop, then the resulting drop due to collector current and $h_{fb} i_b r_c$ (the total rise) is $i_c r_c (1 - \dot{h}_{fb})$. Transforming into the Norton equivalent current generator yields for the generator

$$h_{fb} i_b r_c / (r_c [1 - h_{fb}]) = h_{fe} i_b$$

and $r_c' = r_c (1 - h_{fb})$.

The circuit model for common-emitter connection is shown in Fig. 5.15. The new relation between the input voltage and current is

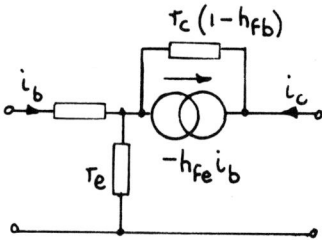

Figure 5.15

derived by noting that

$$v_{be} = -r_e i_e = r_e \cdot \frac{i_b}{1 - h_{fb}}$$

The input resistance of the T then becomes

$$r_T = \frac{v_{be}}{i_b} = \frac{r_e}{1 - h_{fb}} = (h_{fe} + 1) r_e \simeq h_{fe} r_e$$

Since $I_c = h_{fb} I_e \simeq i_e$ it follows that

$$r_e \simeq h_{fe} \cdot \frac{25}{I_c \text{(in mA)}}$$

Because it is often useful to define the output current in terms of an input voltage, we might usefully introduce at this point a transconductance parameter g_m for the bipolar transistor as we did for the FET earlier. Since

$$v_{be} = r_T i_b = h_{fe} r_e i_b$$

$$h_{fe} i_b = \frac{1}{r_e} v_{be} = g_m v_{be}$$

where g_m is the small-signal transconductance. Hence

$$g_m = \frac{1}{r_e} \simeq \frac{I_E (\text{in mA})}{25} = 40 I_E \ \text{mS}$$

Since $I_C \simeq I_E$

$$g_m \simeq 40 I_C \ \text{mS} \qquad\qquad (5.9)$$

This is a transistor parameter which does not vary from one transistor to another but depends only upon the collector current.

In many low-frequency applications, a sufficiently adequate model
consists simply of an input resistance r_T and a controlled current
source $h_{fe} i_b$ or $g_m v_{be}$, see Fig. 5.16.

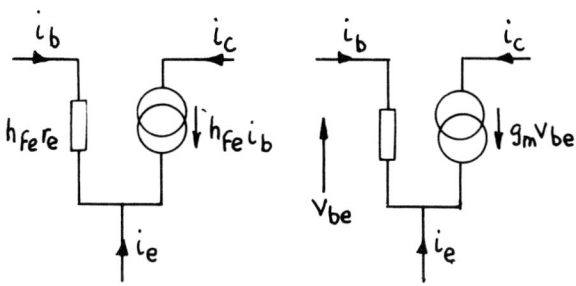

Figure 5.16

<u>Example 5.6</u>   A transistor in common-emitter mode has $h_{fb} = 0.985$
and operates at room temperature with $I_E = 0.8$ mA. For an output
signal of 2.5 V rms developed across a load resistance of 4.7 kΩ,
find the required input signal current and the voltage gain. If the
transistor is now reconnected in common-emitter mode, all other
factors remaining unchanged, calculate the new input signal current
level required and the effective current gain.

Fig. 5.17(a) shows the basic amplifier and the equivalent circuit for common-base connection.

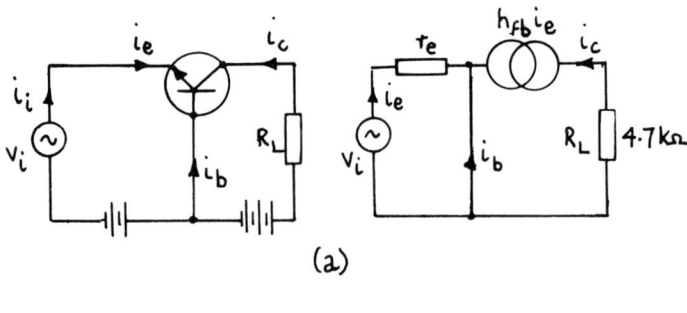

(a)

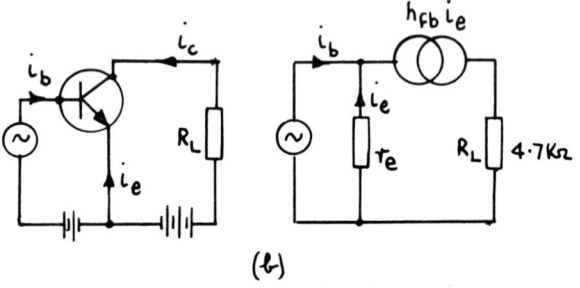

(b)

Figure 5.17

Now

$$r_e = \frac{25}{I_E(\text{in mA})} = \frac{25}{0.8} = 31.25 \ \Omega$$

From the model, input current $i_i = i_e$ and $v_o = h_{fb}i_eR_L$.

$$i_e = \frac{v_o}{h_{fe}R_L} = \frac{2.5}{0.985 \times 4700} = 540 \ \mu\text{A}$$

The input required is $v_i = r_ei_i = 31.25 \times 540 \times 10^{-6}$

$$= 0.0168 \ \text{V}$$

Therefore

$$A_v = \frac{2.5}{0.0168} = 149$$

The common-emitter circuits are shown in Fig. 5.17(b). Then

$$i_e = h_{fb}i_e - i_b \quad \text{or} \quad i_e = -i_b/(1 - h_{fb})$$

$$v_o = h_{fe} i_c R_L = - \frac{h_{fb}}{1 - h_{fb}} \cdot i_b R_L$$

Now

$$i_i = i_b = - \frac{(1 - h_{fb}) v_o}{h_{fb} R_L} = - \frac{0.015 \times 2.5}{0.985 \times 4700}$$

$$= -8.1 \ \mu A$$

This current is considerably less than the 540 μA signal required in the common-base case. There is an effective current gain in the common-emitter connection of $-540/8.1 = -66.7$.

5.7   VOLTAGE, CURRENT AND POWER GAINS

Equations can be derived which often abbreviate calculations when one or more of the gains are known from the use of h-parameters or T-networks. Consider again a four-terminal network with an output load R_L. The defining equations for R_i, A_v and A_i are

$$R_i = \frac{v_1}{i_1} \ ; \quad A_v = \frac{v_2}{v_1} \ ; \quad A_i = \frac{i_2}{i_1}$$

Now

$$A_i = \frac{i_2}{v_2} \cdot \frac{v_2}{v_1} \cdot \frac{v_1}{i_1} = \frac{v_2/v_1 \cdot v_1/i_1}{v_2/i_2}$$

But $v_2 = -i_2 R_L$

Therefore

$$A_v = -A_i \cdot \frac{R_L}{R_i} \tag{5.10}$$

$$A_i = -A_v \cdot \frac{R_i}{R_L} \tag{5.11}$$

Power gain $A_p = |A_i A_v|$ and using equations (5.10) and (5.11) gives us

$$A_p = \left| \frac{A_v^2 R_i}{R_L} \right| = \left| \frac{A_i^2 R_L}{R_i} \right| \tag{5.12}$$

Hence

$$10 \lg A_p = 10 \lg \left[\frac{A_v^2 R_i}{R_L}\right]$$

$$= 20 \lg A_v + 10 \lg \left[\frac{R_i}{R_L}\right] dB$$

5.7.1 Gain curves

The current gain $A_i = h_f/(h_o R_L + 1)$. This equation shows that as $R_L \to 0$, $A_i \to h_f$, and that as $R_L \to \infty$, $A_i \to 0$. $A_i = \frac{1}{2}h_f$ when $R_L = 1/h_o$, and the current gain is therefore 6 dB down on its maximum limiting value of h_f at this value of R_L. If a plot of A_i against $\lg R_L$ is made, the result is as shown in Fig. 5.18. The shape of the curve does not depend on the transistor or the circuit configuration but the absolute magnitude does.

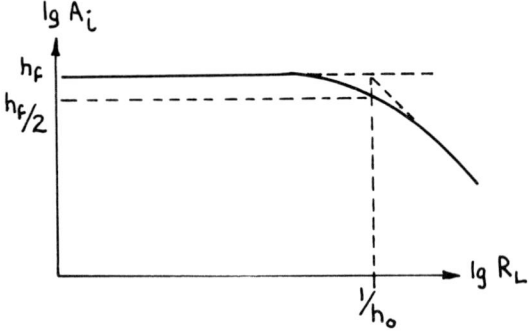

Figure 5.18

Figure 5.19 shows the plots of A_{ib}, A_{ie} and A_{ic} against $\lg R_L$. Notice that A_{ib} and A_{ic} are negative as the conversion formulae indicated.

Turning to voltage gain where $A_v = -h_f R_L/(h_i + \Delta R_L)$ this equation shows that A_v is 6 dB down when $R_L = h_i/\Delta$, for dividing top and bottom by R_L gives

$$A_v = -\frac{h_f}{h_i/R_L + \Delta}$$

and when $R_L \to 0$, $A_v \to 0$, and when $R_L \to \infty$, $A_v \to h_f/\Delta$. Hence, for $R_L = h_i/\Delta$, $A_v = h_f/2\Delta$, that is, 6 dB down on the limiting maximum

158

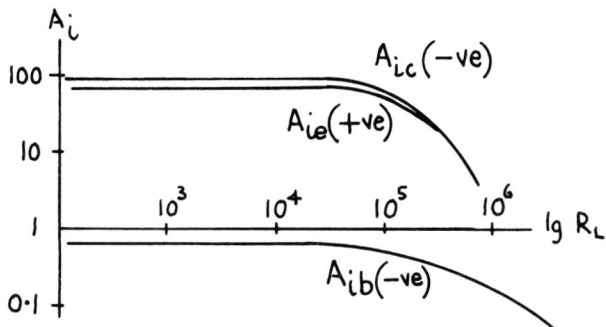

Figure 5.19

of h_f/Δ.

A graph of A_v against R_L is shown for the three configurations, Fig. 5.20. The type of variation is similar to that for A_i provided R_L is replaced by its reciprocal, and on a log scale this simply reverses the graph with respect to that axis. Notice that A_{vb} and A_{vc} give no phase change while A_{ve} gives 180° reversal.

5.8   JUGFET EQUIVALENT CIRCUIT

Because the input resistance of the FET, like the thermionic valve, is very high, the gate current is negligible and the small-signal model is of similar form to that of the thermionic valve. The gate and drain current functions may therefore be expressed as

$$i_g = 0$$

$$i_d = f(v_{GS}, v_{DS})$$

Clearly, only the second of these functions is of significance. The total differential drain current is then

$$di_D = \frac{\partial i_D}{\partial v_{GS}} \cdot dv_{GS} + \frac{\partial i_D}{\partial v_{DS}} \cdot dv_{DS} \tag{5.13}$$

that is, a change in the gate voltage causes a change in drain current which in turn causes a further change in drain current separate from but simultaneous with the original change.

But with v_{DS} held constant, $dv_{DS} = 0$ and the partial derivative

159

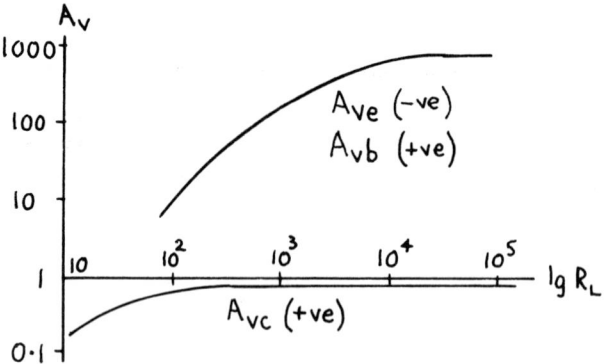

Figure 5.20

with respect to v_{GS} is equal to the total derivative or

$$\frac{\partial i_D}{\partial v_{GS}} = \frac{di_D}{dv_{GS}}\bigg|_{v_{DS} = k} = g_m$$

where g_m is the mutual or transconductance in Siemen. This parameter, relating the dependence of drain current i_D to the gate-source volt-age v_{GS}, represents the gradient of the transfer characteristic of a FET.

A second parameter can be defined by holding v_{GS} constant so that $dv_{GS} = 0$ and then

$$\frac{\partial i_D}{\partial v_{DS}} = \frac{di_D}{dv_{DS}}\bigg|_{v_{GS} = k} = \frac{1}{r_d}$$

where r_d is the drain slope resistance. This parameter, relating the dependence of drain current i_D to the drain-source voltage v_{DS}, re-presents the reciprocal of the gradient of the output characteristic in the saturation region.

Replacing the differentials of (5.13) above by small signals and using the FET parameters g_m and r_d, we have

$$i_D = g_m \cdot v_{GS} + \frac{1}{r_d} \cdot v_{DS} \tag{5.14}$$

and this equation must be satisfied by the a.c. circuit of an equi-

160

valent model. The drain current then consists of two parts and a parallel circuit is required: one branch of this circuit contains a controlled current source $g_m \cdot v_{GS}$ directly proportional to the input signal voltage, the other branch carries a current v_{DS}/r_d directly proportional to the output voltage. The circuit of Fig. 5.21 then follows, the drain load resistance R_L being effectively across the output terminals. From the diagram

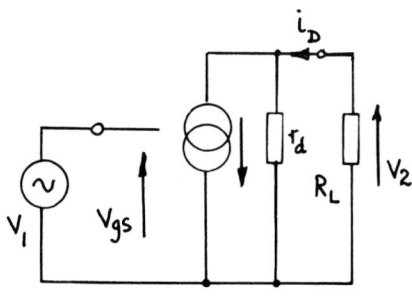

Figure 5.21

$$v_2 = -g_m \cdot v_1 \frac{r_d R_L}{r_d + R_L}$$

and the voltage gain follows at once:

$$A_v = \frac{v_2}{v_1} = -g_m \cdot \frac{r_d R_L}{r_d + R_L} \tag{5.15}$$

For $r_d \gg R_L$ (as is the usual case), $A_v \simeq -g_m R_L$. Reasonably good assessments of performance can be made with an equivalent circuit consisting of a controlled current source $g_m \cdot v_1$ only.

Essentially we have chosen the basic y-parameter circuit discussed under Section 5.2.2 as our FET equivalent model. The input admittance y_i is negligibly small and has been taken as zero, as has the parameter y_r taken as the ratio $\delta i_G / \delta v_{DS}$ with v_{GS} constant. The parameter y_f is the forward transconductance and is equivalent to g_m. Finally the parameter y_o is the output admittance $1/r_d$. The y equivalent circuit and its simplification is shown in Fig. 5.22(a) and (b).

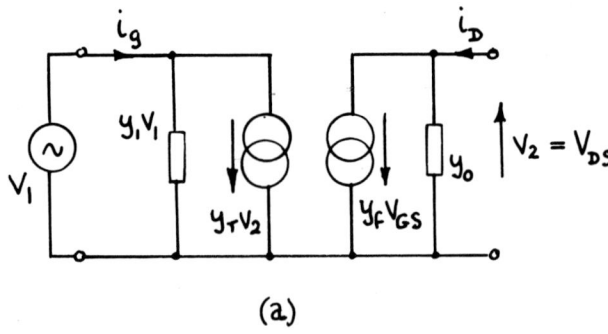

(a)

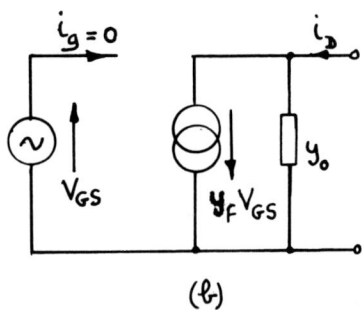

(b)

Figure 5.22

5.8.1 Transconductance

For the JUGFET or the depletion MOSFET, the drain current in the normal operating region has already been mentioned in Chapter 2, equation (2.5):

$$i_D = I_{DSS} \left[1 - \frac{v_{GS}}{V_p} \right]^2 \tag{5.16}$$

The gradient of the transfer characteristic, di_D/dv_{GS}, is the mutual conductance g_m, so by differentiating

$$g_m = \frac{di_D}{dv_{GS}} = -\frac{2I_{DSS}}{V_p} \left[1 - \frac{v_{GS}}{V_p} \right] \text{ Siemen}$$

When $V_{GS} = 0$, $g_{mo} = -2I_{DSS}/V_p$ and so

$$g_m = g_{mo} \left[1 - \frac{v_{GS}}{V_p} \right] = g_{mo} \sqrt{\frac{I_D}{I_{DSS}}} \tag{5.17}$$

162

since, from (5.16)

$$\frac{V_{GS}}{V_p} = 1 - \sqrt{\frac{I_D}{I_{DSS}}}$$

Equation (5.17) is useful in that it enables g_m to be calculated for any operating value of I_D or V_{GS}. As the manufacturers usually state the value of I_{DSS} and g_{mo} the calculation is very easy.

Example 5.7   The amplifier of Fig. 5.23 uses a junction gate FET having the following parameters: V_p = -2.2 V, I_{DSS} = 2.5 mA. It is required to bias the FET at the point I_D = 1.2 mA. If the drain supply V_{DD} = 12 V, obtain estimations of the required gate bias, the mutual conductance, the source resistor R_s and the required value of R_L to obtain a voltage gain of 12. What is the quiescent drain voltage?

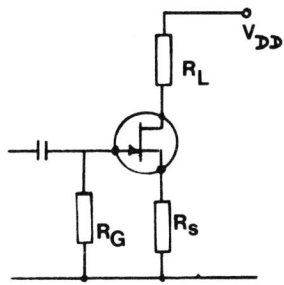

Figure 5.23

From equation (5.16) by rearrangement

$$V_{GS} = V_p \left[1 - \sqrt{\frac{I_D}{I_{DSS}}} \right]$$

and inserting values

$$V_{GS} = -2.2 \left[1 - \sqrt{\frac{1.2}{2.5}} \right] = - 0.675 \text{ V}$$

This voltage has to be dropped across R_s to provide the gate bias. Hence, for a source current I_s (= I_D) of 1.2 mA

$$R_s = \frac{0.675 \times 10^3}{1.2} \simeq 560\,\Omega$$

From equation (5.16) after differentiation

$$g_m = -\frac{2 \times 2.5}{-2.2}\left[1 + \frac{0.675}{2.2}\right] \simeq 3mS$$

Assuming $r_d \gg R_L$, we have the stage gain given by

$$A_v \simeq -g_m R_L$$

hence for a gain of 12 with $g_m = 3mS$, $R_L = 4\ k\Omega$.

The quiescent drain voltage is $12 - (4 \times 1.2) = 7.2$ V.

5.9   HIGH FREQUENCY MODELS

So far we have considered equivalent models by assuming that the operating frequency is not high enough for the charging currents associated with the internal capacitances of the transistors to be significant. Although the operation of transistors is modified at high frequencies by the presence of such capacitances (and other parameters), the main discussion about this aspect will be left to Section 5.9.1. Here we shall briefly mention the form of the equivalent FET and bipolar high frequency models.

The FET is fundamentally rather simpler to analyse than is the bipolar transistor. In Fig. 5.24(a), the small signal model of Fig. 5.21 earlier has been modified to take into account the internal gate-to-source capicitance C_{gs} and the gate-to-drain capacitance C_{gd}. In the JUGFET these capacitances arise, of course, from the reverse biased junctions and will be of the order of 1 to 10 pF or so. In general, C_{gd} is smaller than C_{gs}, but its presence is highly significant. The output capacitance C_{ds} can, in most cases, be ignored.

The main problem is that the effective input capacitance is not simply C_{gs} as it might at first seem, but C_{gd} multiplied (approximately) by the gain of the amplifier. This additional input capacitance arises from what is known as the Miller effect. Feedback takes place through C_{gd} and modifies the input impedance in a manner depending upon the nature of the drain load. This will also

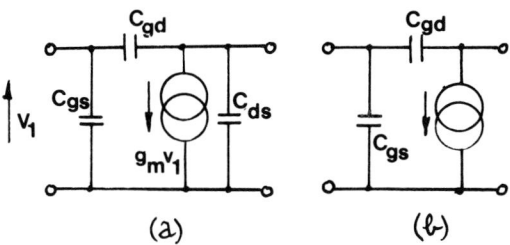

Figure 5.24

be discussed in due course. The simple circuit of Fig. 5.24(b) is usually satisfactory for any but the most precise calculations on high frequency performance of the FET.

For the bipolar transistor, the normal hybrid circuit of Fig. 5.5 can be readily modified to take account of high frequency effects, so that the model of Fig. 5.25 results. This is known as the hybrid-π circuit. The first step is to separate the input resistance into two parts; the first part r_b accounts for that portion of the base resistance associated with the base lead connection and that part of the base that does not lie in the active region between the emitter and collector. The model is completed by adding the capacitors C_e and C_j. Capacitor C_e is the emitter storage capacitance and takes care of the diffusion and transit-time effects at the base-emitter junction; capacitor C_j is the depletion region capacitance at the base-collector junction.

The point of importance is that the controlled current source depends upon the current through the capacitance-shunted r_b rather than the input current i_1. At low frequencies where the reactances of C_e and C_j are very large, $i_1 = i$. At high frequencies, however, $i_i \neq i$. In particular, if the input current is of constant amplitude and sinusoidal, the amplitude of i is a function of frequency. The low frequency gain of the transistor h_{FE} which is closely equal to h_{fe} at low and moderate frequencies, therefore falls as high frequencies are reached. The same is true of h_{fb} in the common-base configuration. The frequency at which either $|h_{fe}|$ or $|h_{fb}|$ drops to $1/\sqrt{2}$ of its low frequency value is known as the 'alpha cutoff

165

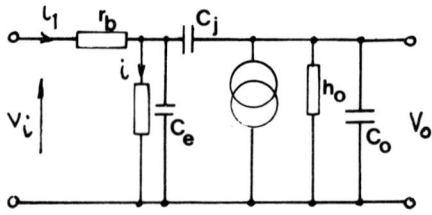

Figure 5.25

frequency', symbolised $f_{\alpha e}$ or $f_{\alpha b}$ respectively.

5.9.1 Analysis of high frequency performance

Ignoring the distributed nature of the diffusion process inside the transistor, the frequency dependence of h_{fe} can be expressed to a close approximation by the equation

$$h_{fe} = \frac{h_{FE}}{1 + j\dfrac{f}{f_{\alpha e}}} \qquad (5.18)$$

where $f_{\alpha e}$ is the common-emitter alpha cutoff frequency defined above. From this equation the modulus of h_{fe} is then

$$|h_{fe}| = \frac{h_{FE}}{\sqrt{1 + [\dfrac{f}{f_{\alpha e}}]^2}} \qquad (5.19)$$

and its phase angle is

$$\underline{/\theta}_e = -\tan^{-1}\left[\frac{f}{f_{\alpha e}}\right]$$

Now from basic theory

$$h_{fb} = \frac{h_{fe}}{1 + h_{fe}}$$

hence from (5.18)

$$h_{fb} = \frac{h_{FE}}{1 + j\dfrac{f}{f_{\alpha e}}} \div 1 + \frac{h_{FE}}{1 + j\dfrac{f}{f_{\alpha e}}}$$

which reduces to

$$h_{fb} = \frac{h_{FE}}{1 + h_{FE} + j\dfrac{f}{f_{\alpha e}}}$$

Dividing top and bottom by $(1 + h_{FE})$ yields

$$h_{fb} = \frac{h_{FE}}{1 + h_{FE}} \cdot \left[1 + \frac{jf}{f_{\alpha r}(1 + h_{FE})}\right]^{-1}$$

$$= \frac{h_{FB}}{1 + j\dfrac{f}{f_{\alpha b}}} \tag{5.20}$$

It then follows that the modulus of h_{fb} is

$$|h_{fb}| = \frac{h_{FB}}{\sqrt{1 + [\dfrac{f}{f_{\alpha b}}]^2}} \tag{5.21}$$

with phase angle

$$\underline{/\theta_b} = -\tan^{-1}\left[\frac{f}{f_{\alpha b}}\right] \tag{5.22}$$

The common-base cut-off frequency $f_{\alpha b}$ is clearly

$$f_{\alpha b} = f_{\alpha e}(1 + h_{FE}) \tag{5.23}$$

From this result, since

$$\frac{1}{1 + h_{FE}} = 1 - h_{FB}$$

$$f_{\alpha e} = f_{\alpha b}(1 - h_{FB}) \tag{5.24}$$

It is useful to notice that the product $h_{FE}f_{\alpha e}$ = the product $h_{FB}f_{\alpha b}$.

It follows from equations (5.23) and (5.24) that the alpha cut-off frequency in common-base connection is very much greater than that in common-emitter connection.

Example 5.8   A transistor has the following common-base parameters: h_{FB} = 0.98, $f_{\alpha b}$ = 15 mHz. Sketch curves, clearly showing the scale values and cutoff frequencies, of the frequency dependence of the modulus and phase angle of each of h_{fb} and h_{fe}. What conclusion can be drawn as to the relative merits of the two modes of connection for operation at high frequencies?

From equation (5.21), inserting values, we have

$$h_{fb} = \frac{0.98}{\sqrt{1 + [\frac{f}{15}]^2}}$$

Taking a range of frequency values in the range .01-100 MHz gives the graph as sketched in Fig. 5.26. Also

$$\underline{/\theta_b} = -\tan^{-1}\frac{f}{15}$$

and this leads to the phase variation plot for the common-base connection.

For common-emitter

$$h_{fe} = \frac{0.98}{1 - 0.98} = 49$$

and hence

$$f_{\alpha e} = 15(1 - 0.98) = 0.3 \text{ MHz}$$

Therefore

$$|h_{fe}| = \frac{49}{\sqrt{1 + [\frac{f}{0.3}]^2}}$$

Also

$$\underline{/\theta_e} -\tan^{-1}\frac{f}{0.3}$$

The plots for the modulus and phase angle of the common-emitter are

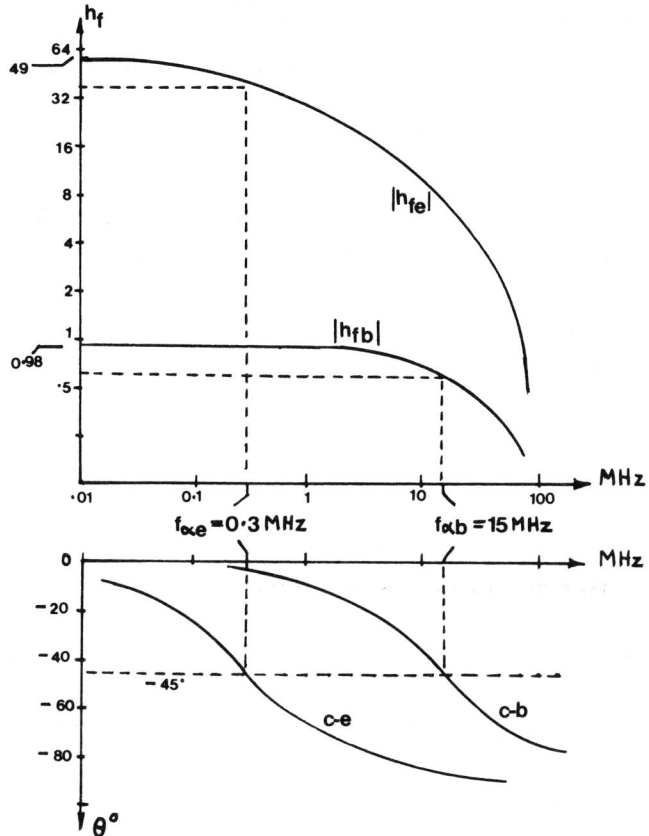

Figure 5.26

then readily drawn.

Clearly, the common-base connection makes the most effective use of the transistor's useful frequency range.

PROBLEMS 5

1. What do you understand by the term 'small signal'?

2. A diode has a static resistance of 10 Ω and a dynamic resistance of 1 kΩ. Explain what is meant by this statement.

3. Why are small-signal parameters not used for the analysis of large-signal amplifiers?

169

4. Why is the battery (or the d.c. supply) to an amplifier not shown on the equivalent model? How is such a supply actually represented?

5. Define the h-parameters. Express the common-emitter h-parameters in terms of the common-base parameters.

6. Determine the z-parameters of the circuit shown in Fig. 5.27.

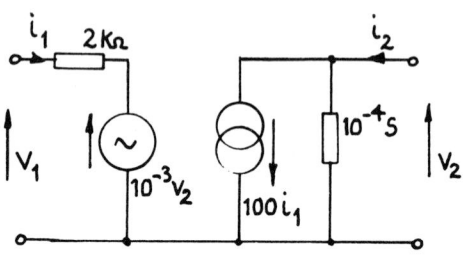

Figure 5.27

7. Define a set of h-parameters for a transistor and explain, by setting down the initial equations, how these parameters can be used to determine the small-signal gain of the transistor when inserted between a specified source and a load.

8. As h_{oe} and h_{re} are usually very small, they are often neglected in calculations. Show that this leads to the following approximations:

$$A_i \simeq h_{fe}; \quad A_v \simeq - h_{fe}R_L/h_{ie}; \quad A_p \simeq h_{fe}^2 R_L/h_{ie}$$

9. Sketch a transfer and an output characteristic for a junction-gate FET and define from your graphs the parameters g_m, r_d and μ.

10. Given the following functional relationships

$$i_1 = f(v_1, v_2)$$

$$i_2 = f(v_1, v_2)$$

and assuming a non-linear two-port network, derive a small-signal model with admittance parameters y_i, y_r, y_f and y_o.

11. A semiconductor two-port network is described by the equations

$$i_1 = 10^{-6} \exp (40v_1)$$

$$i_2 = -0.99i_1 + 10^{-6}v_2$$

Sketch a small-signal model for this circuit at an operating point defined by $i_1 = 1.5$ mA and $v_2 = -6$ V.

12. A transistor is connected as a common-base amplifier with a collector load of 10 kΩ. If the appropriate h-parameters are $h_{ib} = 30$ Ω, $h_{fb} = -0.98$, $h_{ob} = 2 \times 10^{-6}$ S, $h_{rb} = 3 \times 10^{-4}$, sketch an equivalent circuit and calculate (a) the input resistance, (b) the voltage gain.

13. The small-signal voltages and currents for the transistor shown in Fig. 5.28 are defined by the following equations:

$$v_1 = 10^3 i_1 + (6 \times 10^{-4})v_2$$

$$i_2 = 50i_1 + 10^{-4}v_2$$

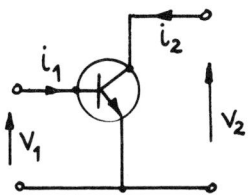

Figure 5.28

Identify the four hybrid parameters and state the units, if any, in which they are measured.

The transistor is connected so that the bias conditions are unchanged to form a small-signal amplifier having a collector load of 20 kΩ and fed from a source of negligible internal resistance. Calculate the current, voltage and power gains of the amplifier, expressing this last in dB.

14. You have a transistor with the following parameters: $h_{ie} = 2$ kΩ,

h_{fe} = 50, h_{oe} = 5 × 10^{-4}S, h_{re} being negligible. This transistor is
to be used as a simple pre-amplifier connecting the output from a
pick-up to the main amplifier. The output from the pick-up (which
has negligible internal impedance) is 1.5 mV, but 0.15 V is required
as input to the main amplifier to provide full output. Assuming that
the bias conditions are correct, what value of load resistor would
you select for the pre-amplifier collector?

15. Sketch an equivalent circuit for a JFET common-source amplifier.
What are the defining equations for this circuit? Using the equivalent circuit, show that the voltage gain of the amplifier is
approximately $-g_m R_L$. What assumptions have you made in obtaining
this expression?

16. A JFET has g_{mo} = 3mS and V_p = -4 V. Estimate its mutual conductance g_m at (a) V_{GS} = -1 V, (b) I_D = 1.5 mA.

17. A depletion MOSFET has I_{DSS} = 8 mA and g_{mo} = 4 mS. Estimate g_m
at (a) I_D = 6 mA, (b) V_{GS} = -1 V.

18. The JFET of Problem 16 is used in the circuit of Fig. 5.29,
where the gate bias is set to -1 V. Predict the quiescent value of
the drain current. What input signal amplitude is required to produce an output voltage of 50 mV? What 'practical' value of source
resistance R_s would you use?

19. The following values were taken from the linear parts of the
static characteristics of a certain JFET:

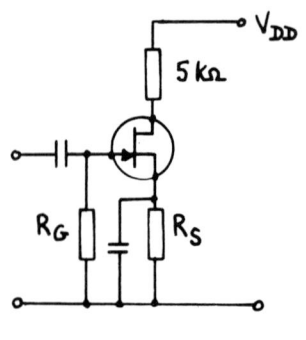

Figure 5.29

V_{DS} (V)	12	12	8
I_D (mA)	7.8	5.2	7.4
V_{GS} (V)	-1	-2	-1

Estimate the parameters g_m, r_d and μ.

20. A JFET obeys the square law relationship of Equation (5.16) and has I_D = 10 mA when V_{GS} = 0. If V_p for this transistor is -5V, find I_D and g_m at V_{GS} = 0, 1, 2, 3, 4 and 5 V, and plot the transfer characteristic. Estimate the value of g_m when I_D = 1.5 mA.

21. A common-emitter amplifier uses a transistor with the following parameters: h_{ie} = 15 kΩ, h_{fe} = 90, h_{re} = 5 × 10^{-4}, h_{oe} = 20 uS. If the amplifier is to have an input resistance of 1.2 kΩ, find the value of load resistor to be employed. Under these conditions calculate A_v and A_i.

6 Small signal amplifiers

6.1   INTRODUCTION
Because the signal excursions are small in comparison to the d.c.
bias and the operating point values of current and voltage, and Class-
A operation is concerned, the actual bias of small-signal amplifying
stages is not normally critical and the distortion which arises in
large-signal amplifiers is easily avoided. Such small signal stages
are usually best treated by the application of linear equivalent
models, so avoiding the tedious work so often associated with
graphical analysis.

A.C. COUPLING
Those simple single-stage untuned amplifiers discussed in terms of
equivalent circuits in the previous chapter have not been considered
in the light of their performance over a range of frequencies but
have rather been assumed to be operating at which might be called
'moderate' frequencies. That is, consideration of what happens at
very low frequencies, say from 100 Hz downwards, or at high frequencies, say from a few hundred kilohertz upwards, have been
avoided. We shall now consider the operation of untuned amplifiers
over an extended range of frequencies and look into the effects of
cascading such stages. At this point we shall consider both bipolar and field-effect transistors in a.c. coupled stages, leaving
the discussion of zero frequency or d.c. amplifiers until a little
later on.

6.2.1   Resistance-capacitance coupling
Fig. 6.1(a) and (b) show respectively the general configuration for

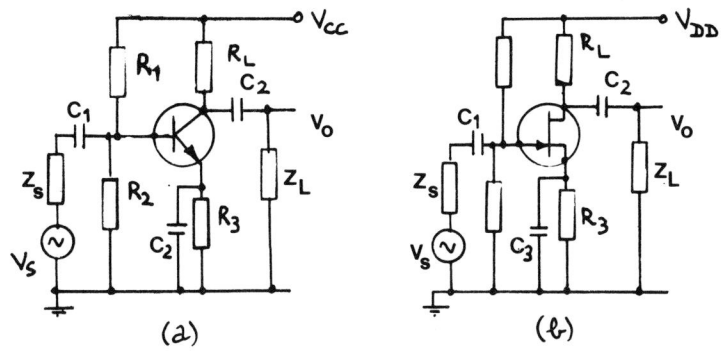

Figure 6.1

a single-stage common-emitter and common-source amplifier. The signal input is assumed to come from a Thevenin equivalent source having an open-circuit e.m.f. of V_S volts and internal impedance Z_S. The output load Z_L will, in most cases, be the input impedance of a following stage of amplification. Both input and output signals are coupled to their respective loads by capacitors C_1 and C_2 and the associated resistances, hence the name 'resistance-capacitance' coupling. The gain of the amplifiers is A and their input and output impedances are respectively Z_i and Z_o. In general, A, Z_i and Z_o are complex functions of frequency.

We are not concerned in analysis with the d.c. bias and operating points of the amplifiers; these are taken to be correctly adjusted. Bias stability resistors R_3 are in both cases short-circuited to a.c. by capacitors C_3 and the d.c. supplies V_{cc} and V_{DD} are effectively at zero a.c. impedance. There are, however, certain elements present in the circuits which do not appear on the circuit diagrams. The transistors have small but often significant internal capacitance and there is stray capacitance in the wiring and between component parts. Taking these points into account, the equivalent circuits of Fig. 6.2(a) and (b) can be drawn for the two amplifiers, assuming that the source and load impedances are both resistive.

In diagram (a) the input resistance of the transistor is given as h_{ie}; this is justified because as we have seen previously the term $h_{fe}h_{re}/(h_{oe} + 1/R_L)$ which appears in the input impedance

175

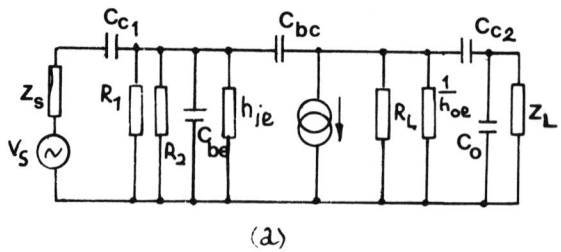

(a)

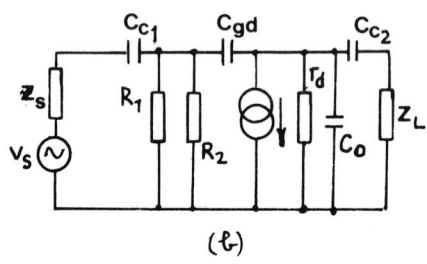

(b)

Figure 6.2

equation which precedes equation 5.7 is usually negligible in comparison with h_{ie}. Resistors R_1 and R_2 are effectively in parallel with h_{ie} as is C_{be}, the input capacitance of the transistor. The base-collector capacitance C_{bc} connects input and output terminals. The current source $h_{fe}i_i$ feeds into parallel resistors R_L and $1/h_{oe}$ and shunt capacitance C_o. The FET equivalent circuit of Fig. 6.2(b) may be followed through in the same manner.

Both these equivalent circuits are complicated and the evaluation of the individual elements is not easy if at all possible in a practical case since they vary with the d.c. conditions, the configuration, the layout and such other factors as temperature and the characteristics of the preceding or following stages. It is fortunate that the circuits can be simplified a bit at a time, as it were, if specific frequency ranges are considered in turn; these ranges are conveniently specified as low, intermediate (or mid-frequency) and high. They are relative to the overall frequency coverage of the amplifier in question and are not identical for all amplifiers. What might be the mid-frequency range of an audio-amplifier, for example, might well be considered as the low-frequency range of a video amplifier.

6.2.2   Coupling response

As both model circuits are similar in form, the same analysis may be used for both.

If we examine the voltage gain relative to frequency for a single stage RC coupled amplifier, such as either of those illustrated in Fig. 6.1, the result would be something like that shown in Fig. 6.3. Notice the logarithmic division of the frequency axis and the specific frequency divisions. The gain is substantially constant over the mid-frequency range but falls off at both the low- and high-frequency ranges. The bandwidth of the amplifier is defined as that span of frequencies over which the power gain has not fallen by more than 3 dB from the mid-frequency gain figure A_o. The limiting frequencies f_1 and f_2 are the half-power points and at these points the response is $1/\sqrt{2} = 0.707$ of A_o.

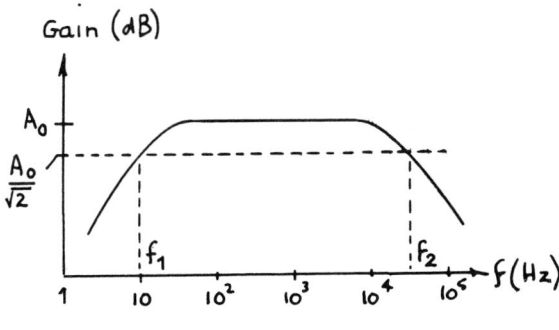

Figure 6.3

Ignoring wiring inductance which is generally negligible at frequencies up to a few megahertz and the behaviour of the active devices themselves at very high frequencies, the only components in the circuits that are frequency sensitive are the capacitors, coupling and stray. The fact that there is a substantial frequency band in which the gain is constant suggests that the capacitive effects are negligible over this band. Considering series and parallel combinations of resistance and capacitance we have, referring to Fig. 6.4

$$Z_s = R - jX_c$$

Therefore

$$|z_s| = \sqrt{R^2 + X_c^2} = R\sqrt{1 + 1/(\omega CR)^2} \qquad (6.1)$$

$$Z_p = \frac{1}{1/R + jXc}$$

Therefore

$$|z_p| = \frac{1}{1/R^2 + 1/X_c^2} = \frac{R}{\sqrt{1 + (\omega CR)^2}} \qquad (6.2)$$

From (6.1) if $\omega CR \gg 1$, $Z_s \simeq R$; and from (6.2) if $\omega CR \ll 1$, $Z_p \simeq R$. Since ωC increases with frequency, the series capacitance of Fig. 6.4 becomes important only at frequencies within the low frequency range, and the parallel capacitance becomes important only at frequencies within the high frequency range.

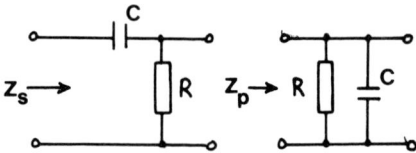

Figure 6.4

6.3.   MODELS FOR THE FULL FREQUENCY RANGE

6.3.1   The mid-frequency model

The fact that the mid-frequency gain is constant suggests as already noted, that the effect of both coupling and stray capacitance may be neglected. The equivalent circuits of Fig. 6.2(a) and (b) may therefore be simplified to the purely resistive models of Fig. 6.5(a) and (b).

For the bipolar amplifier at (a), R_B which is the equivalent of bias resistors R_1 and R_2 in parallel, must be small compared with $\alpha_E R_E$ ($\simeq h_{FE} R_E$), from equation (3.8), and for insignificant loading of the input R_B should be large compared with h_{ie}. So we assume that $h_{fe} R_e \gg R_B \gg h_{ie}$; the circuit will then be stable and the analysis relatively easy. In the output circuit $1/h_{oe}$ is in parallel with R_L. Taking $1/h_{oe}$ to be large in comparison with R_L the output voltage

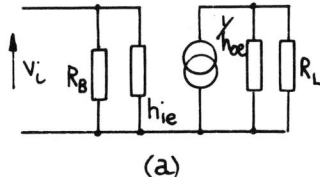 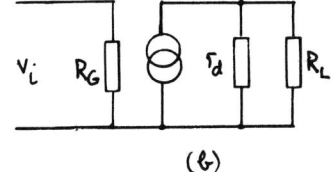

(a)   (b)

Figure 6.5

$$v_o \simeq -h_{fe} i_i R_L \simeq - \frac{h_{fe} v_i R_L}{h_{ie}}$$

Hence the mid-frequency voltage gain is

$$A_{vo} \simeq - \frac{h_{fe} R_L}{h_{ie}}$$

For the FET model, since the input resistance R_G is very large in comparison with R_s, it can be neglected, and since r_d is in general large compared with R_L, the output voltage is

$$v_o \simeq -g_m v_i R_L$$

so that the mid-frequency voltage gain is

$$A_{vo} \simeq -g_m R_L$$

As we should expect, both these results for gain are the same as those obtained in the previous chapter when frequency considerations were ignored. For the phase shift, since capacitances are omitted the coupling shift is zero and the overall shift is that due to the transistor alone; for common-emitter connection this is 180°.

6.3.2   The low-frequency model

In the low-frequency range of the overall spectrum the reactances of the parallel capacitors are extremely high and their shunting effect is negligible. For the series connected coupling capacitors C_{c1} and C_{c2}, however, increasing reactance becomes important. The low frequency equivalent circuits for the bipolar and FET amplifiers become now as shown in Fig. 6.6(a) and (b) respectively.

In both cases the input and output circuits are first-order high-pass filters of the general form shown in Fig. 6.7. Both circuits can be analysed from this diagram. Consider for the moment the effect of the output coupling involving C_{c2} (call this C) and the total effective resistance (call this R_T) where R_T is the sum of the resistances 'facing' the capacitor; that is the resistance encountered in moving around the circuit from one capacitor plate to the other. The gain of the filter at low frequencies is then clearly

$$\frac{V_o}{V_i} = \frac{R_T}{R_T - j\frac{1}{\omega C}}$$

and

$$\left|\frac{V_o}{V_i}\right| = \frac{1}{1 + \frac{1}{(\omega C R_T)^2}} \qquad (6.3)$$

This gain is, of course, less than unity.

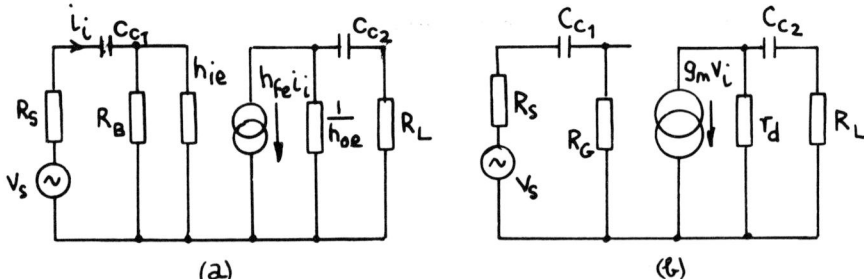

(a)   (b)

Figure 6.6

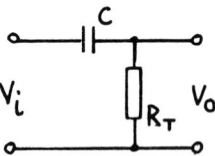

Figure 6.7

The low frequency output of either input or output CR network is therefore related to the mid-frequency output by a complex factor depending upon ω and a CR product. As the frequency falls, a larger fraction of the input voltage is dropped across the appropriate coupling capacitor and the output voltage developed across the associated resistance correspondingly decreases. The half-power frequency is reached in either case when $|v_o/v_i| = 0.707 \, A_o$; if this frequency is

ω_1 then $\omega_1 CR_T = 1/\omega_1$ and $CR_T = 1/\omega_1$.

Substituting for CR_T into (6.3) above yields the low frequency gain figure in terms of the half-power frequency and the general frequency ω:

$$A_L = \left| \frac{v_o}{v_i} \right| = \frac{1}{\sqrt{1 + (\frac{\omega_1}{\omega})^2}} \qquad\qquad (6.4)$$

It is this relationship which leads to the steady decline in the shape of the response curve at low frequencies, see Fig. 6.3. Consider again the output coupling only. If the frequency $\omega \gg \omega_1$, then $A_L = A_o$; if $\omega_1 \gg \omega$, then $A_L \approx \frac{\omega}{\omega_1} A_o$. So the gain is inversely proportional to frequency, and every time the frequency is increased by one octave (doubled) the voltage gain falls by 6 dB (one-half). You should be able to deduce for yourself that this is equivalent to a roll off of 20 dB per decade. Thus the response curve at low frequencies is asymptotic to a gradient of -6 dB per octave and there is no need to plot the curve accurately point by point. The half-power frequency ω_1 comes at the intersection of the -6 dB per octave asymptote and the 0 dB level corresponding to the mid-frequency gain level. The response curve can now be sketched as Fig. 6.8 shows.

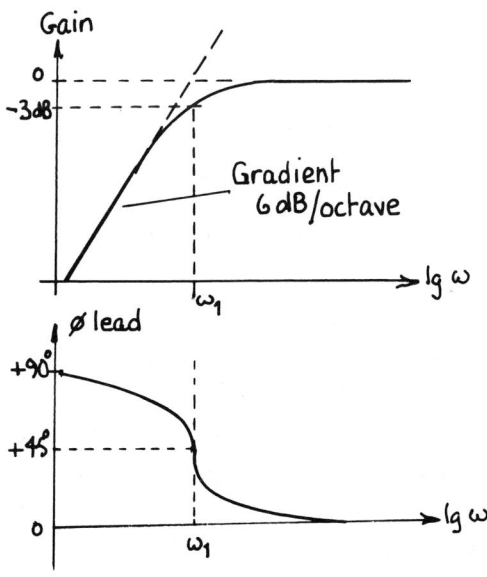

Figure 6.8

181

We now have to consider phase shift. Again, only two points need be considered. The complex gain $A_L = 1/(1 - j\frac{\omega_1}{\omega})$ and for $\omega \ll \omega_1$, $A_L \simeq j\frac{\omega}{\omega_1}$. For phase angle ϕ between v_o and v_i

$$\tan \phi = \frac{\omega/\omega_1}{0} \rightarrow \infty$$

Thus $\phi \rightarrow 90^o$ as the frequency decreases and v_o leads v_i by some angle between 0° and 90° over the low frequency range.

At the 3 dB point, $\omega = \omega_1$, hence $\tan \phi = 1$ and v_o leads v_i by 45°. The phase shift curve is sketched in Fig. 6.8.

Whilst all amplifiers having RC coupled stages inevitably exhibit a declining gain with an associated phase shift, at low frequencies, it is possible to design an amplifier with a flat response down to zero frequency; this characteristic is found in virtually all integrated circuit amplifiers. Such d.c. amplifiers are directly coupled and capacitors are avoided altogether. We shall deal with such systems in later chapters.

6.3.3 The high-frequency model

In the high frequency range of the overall spectrum the reactances of the coupling capacitors are negligible, but the decrease in the reactances of shunt capacitors becomes of importance. The high frequency equivalent circuits of bipolar and FET small-signal stages are then as shown in Fig. 6.9(a) and (b) respectively. Here the input and output shunt capacitances C_i and C_o are taken to be the lumped versions of the device capacitances and wiring strays.

The situation is now further complicated, however, by the fact that the internal capacitances C_{bc} and C_{gd} of the bipolar and FET devices respectively become significant and provide a signal feedback path between output and input. This has a considerable effect on the effective input capacitance of the amplifier. Let us examine the conditions that affect the input capacitance at high frequencies.

First, there is C_{be} itself, the capacitance of the forward biased base-emitter junction. This capacitance arises from two basic causes: firstly there is simply the capacitance created by the depletion layer 'dielectric' which is much narrower in a forward-biased junction than in a reverse-biased one and so has a higher

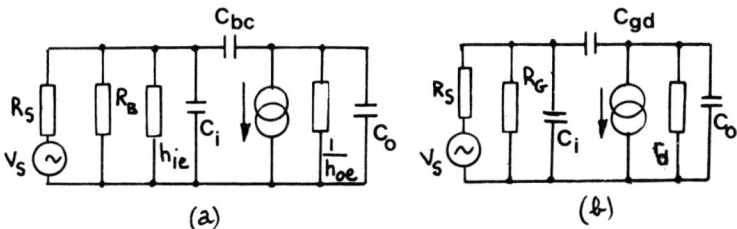

Figure 6.9

capacitance; secondly, additional capacitance arises from the transit time effect of the minority carriers as they diffuse across the junction. These carriers, because of the time spent in the base region, appear to be temporarily stored in the material when the external signal changes rapidly, so as far as the signal is concerned, the effect is similar to charge storage by a conventional capacitor. This is diffusion capacitance already mentioned in Chapter 2. The effective C_{be} of a small signal transistor can be of the order of several hundred picofarad as a result of these effects.

There is now the effect of the collector-base capacitance to be considered. Because C_{bc} (or C_{gd}) is very much smaller than C_{be}, typically a few picofarad, it might appear that its effect on the input capacitance would be negligible, but this is not so. Feedback takes place through the base-collector capacitance and modifies the input impedance in a manner depending upon the nature of the collector load. If the load is resistive, the input impedance is purely capacitive; hence the effective input capacitance is increased above that due to the effects of C_{be} alone. Consider the generalised circuit of Fig. 6.10, where C_F is the capacitance coupling input and output portions of the circuit. Now $I_1 = I_2 + I_3$ and $I_2 = j\omega C_i v_i$. Also $I_3 = -j\omega C_F(v_o - v_i)$.

Now

$$\frac{v_o}{v_i} \simeq -g_m R_L$$

Combining these equations yields

183

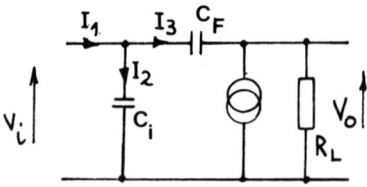

Figure 6.10

$$\frac{v_i}{I_1} = Z_i = \frac{1}{j\omega\{C_i + C_F(1 + g_m R_L)\}}$$

Hence the input capacitance has been increased by an amount $C_F(1 + g_m R_L) \simeq C_F(1 + A_v)$. This is the Miller effect already encountered; the total input capacitance is very much greater than the simple C_i so far considered.

Both circuits of Fig. 6.9 have input and output combinations of resistance and capacitance which represent first order low-pass filters of the general form shown in Fig. 6.11. A basic analysis of the high frequency response of the amplifiers can be made from this filter just as it was in the case of low frequency response. The gain of the filter circuit is

$$\frac{v_o}{v_i} = \frac{1/j\omega C}{1/j\omega C + R_T} = \frac{1}{1 + j\omega C R_T}$$

where again R_T is the total equivalent resistance facing the capacitor. Hence

$$\left|\frac{v_o}{v_i}\right| = \frac{1}{1 + (\omega C R_T)^2} \tag{6.5}$$

and when $|v_o/v_i| = 0.707$, $\omega^2 C^2 R_T^2 = 1$ which is the half-power or -3 dB point. Let this half-power frequency be ω_2 then $C R_T = 1/\omega_2$, and substituting for $C R_T$ into (6.5) above yields the high-frequency gain figure in terms of the half-power frequency and the general frequency ω:

184

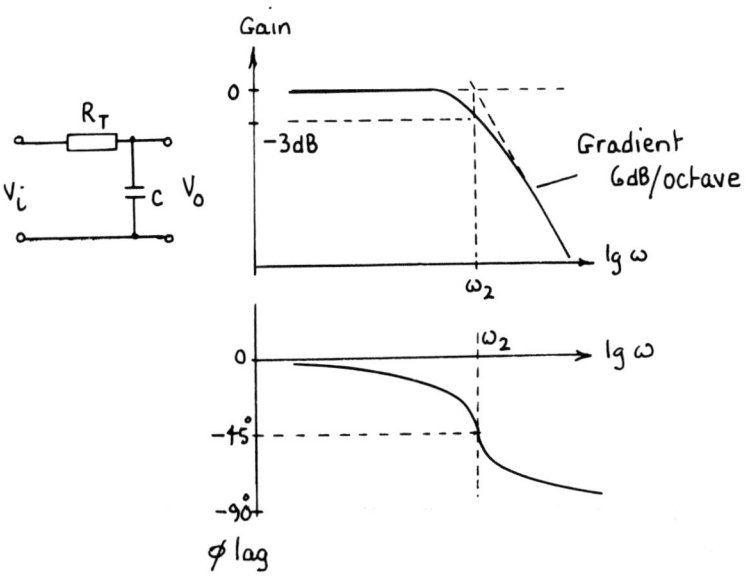

Figure 6.11

$$A_H = \frac{1}{\sqrt{1 + (\frac{\omega}{\omega_2})^2}}$$

The relative gain at high frequencies to that at the mid-frequencies is then

$$|A_H| = \frac{|A_o|}{1 + [\frac{\omega}{\omega_2}]^2} \qquad\qquad (6.6)$$

For $\omega \ll \omega_2$, $A_H \simeq A_o$; for $\omega \gg \omega_2$, $A_H \simeq \frac{\omega_2}{\omega}A_o$. So the gain is inversely proportional to frequency and the response at high frequencies is asymptotic to a gradient of -6 dB per octave. The high frequency response curve can now be sketched as Fig. 6.11 shows.

You can deduce for yourself that the phase angle $\phi \rightarrow 90^o$ as the frequency increases and that the phase angle is -45° at the half-power point. The curve of phase shift is included on Fig. 6.11.

Example 6.1   A FET having g_m = 2.5 mS, r_d = 10 kΩ was used in the circuit of Fig. 6.1(b) with R_s = 500 Ω, R_G = 5 MΩ, R_L = 5 kΩ to provide a mid-frequency gain A_o of -10.   If the equivalent input and output capacitances are 40 pF and 25 pF respectively, determine the upper cut-off frequency.

The equivalent input and output resistances are

$$R_s \| R_G = 500Ω \| 5MΩ \simeq 500Ω$$

$$r_d \| R_L = \frac{10 \times 5}{10 + 5} \, kΩ = 3.34 \, kΩ$$

The individual cut-off frequencies can now be found; for the input coupling let the cut-off frequency be ω_{21}, then

$$\omega_{21} = \frac{1}{C_i (R_s \| R_G)}$$

$$= \frac{1}{40 \times 10^{-12} \times 0.5 \times 10^3}$$

$$= 50 \, Mrad/s$$

Therefore

$$f_{21} = 7.96 \, MHz$$

For the output coupling let the 3dB frequency be ω_{22}, then

$$\omega_{22} = 1/(C_o \{ R_L \| r_d \})$$

$$= 1/(25 \times 10^{-12} \times 3.34 \times 10^3)$$

$$= 11.97 \, Mrad/s \quad or \quad f_{22} = 1.9 \, MHz$$

The upper 3 dB frequency of this amplifier is the figure obtained for f_{22}, that is, 1.9 MHz.   The fact that the input coupling passes frequencies up to 7.96 MHz does not affect the much lower 3 dB frequency characteristic of the output coupling and so the latter determines the effective high frequency response of the amplifier overall.

6.4   THE OVERALL RESPONSE

The previous example has illustrated the point that the overall behaviour at both the low- and high-frequency ends of the spectrum of the bipolar and FET a.c. coupled amplifiers of Fig. 6.1 can be deduced only by taking into consideration the effects of the RC couplings at both input and output ports.

For the low frequency case, referring to Fig. 6.6 and taking $R_B \gg h_{ie}$ in diagram (a), the half-power points where the input voltage will be 70% down on V_s occur for the bipolar and FET circuits respectively at a frequency given by

$$\omega_{11} = 1/(C_{c1}\{R_s + h_{ie}\}) \quad \text{or} \quad 1/(C_{c2}\{R_s + R_G\})$$

and the half-power points where the output will be 70% down on $g_m v_i R_L$ occur respectively at a frequency given by

$$\omega_{12} = 1/(C_{c2}\{R_L + 1/h_{oe}\}) \quad \text{or} \quad 1/(C_{c2}\{R_L + r_d\})$$

The overall gain reduction is then expressed as the product of the individual reductions so that, from equation (6.4), the 'relative' gain at low frequencies for either the bipolar or the FET amplifier is

$$\frac{A_L}{A_o} = \frac{1}{[1 - j\frac{\omega_{11}}{\omega}][1 - j\frac{\omega_{12}}{\omega}]}$$

You should now be able to deduce that the relative gain at high frequencies for either amplifier is

$$\frac{A_H}{A_o} = \frac{1}{[1 - j\frac{\omega}{\omega_{21}}][1 - j\frac{\omega}{\omega_{22}}]}$$

where

$$\omega_{21} = 1/(C_i\{R_s \| h_{ie}\}) \quad \text{or} \quad 1/(C_i\{R_s \| R_G\})$$

and

$$\omega_{22} = 1/(C_o\{R_L \| 1/h_{oe}\}) \quad \text{or} \quad 1/(C_o\{R_L \| r_d\})$$

where C_i is the Miller equivalent capacitance.

To determine the behaviour of a circuit, evaluate ω_{11} and ω_{12}, and ω_{21} and ω_{22}. The HIGHER of ω_{11} or ω_{12} determines the 'lower' half-power frequency; the 'lower' of ω_{21} or ω_{22} determines the 'upper' half-power frequency. Keep in mind that the input capacitance C_i is the effective Miller capacitance in both cases.

6.5   THE NYQUIST PLOT

The two curves of Fig. 6.8 and Fig. 6.11 constitute a useful description of the amplitude and phase performance of a circuit outside of
the mid-frequency range and are often combined in a single diagram
when they are known as a Bode plot. The same information can be
expressed in polar co-ordinates rather than the Cartesian of the Bode
diagram. This kind of plot, known as the Nyquist diagram, is often
of much greater value in assessing the performance of amplifiers,
particularly where feedback is employed. In Fig. 6.12 let O be the
pole (or origin) of polar co-ordinates, OQ being the line of zero
angle and representing in magnitude the mid-frequency gain A_o. Let
the line OP represent in magnitude and direction (angle) the gain
A_H at some other frequency, ω. Then

$$OP = \frac{|A_o|}{1 + (\frac{\omega}{\omega_2})^2}$$

and angle POR = phase angle ϕ.

Now $\dfrac{OP}{OR} = \sqrt{1 + (\frac{\omega}{\omega_2})^2}$ since OR = $A_o / 1 + (\frac{\omega}{\omega_2})^2$, the real part of phasor OP. But

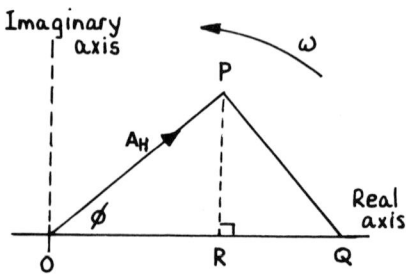

Figure 6.12

$$\frac{OQ}{OP} = \sqrt{1 + (\frac{\omega}{\omega_2})^2}, \quad \text{hence} \quad \frac{OP}{OR} = \frac{OQ}{OP}$$

and the triangles OPR and OPQ are similar. Hence angle OPQ is a right-angle and the locus of P as the frequency increases will be, by elementary geometry, a semicircle. The Nyquist plot for a first-order system is then, as Fig. 6.13 shows, a semicircle of diameter A_0.

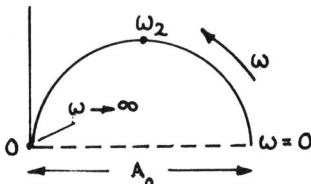

Figure 6.13

Notice that the frequency increases around the locus in the direction indicated. The frequency is equal to zero at the point on the curve most remote from the pole O, passes through the half-power frequency ω_2 at the highest point on the curve (when $\phi = 45^\circ$) and approaches infinity as the curve approaches the origin (when $\phi \rightarrow 90^\circ$ and $A_H \rightarrow 0$). By similar reasoning, which you might try for yourself, the low frequency Nyquist plot simply completes the circle as Fig. 6.14 shows, so providing a full polar plot of the a.c. coupled amplifier.

6.6   BYPASS CAPACITORS

In the analysis of the frequency response, it has been assumed that

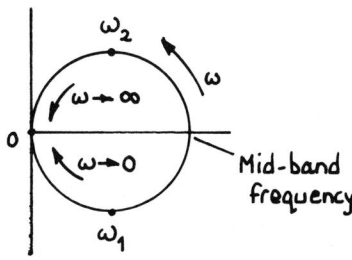

Figure 6.14

the emitter (or source) has been at earth potential from the signal viewpoint, that is, any emitter (or source) resistance has been effectively bypassed by a capacitor of negligible reactance at the lowest frequency of interest. In most cases it is not the coupling capacitors that are the important elements in determining the lower cut-off frequency but the emitter or source bypass capacitor, C_E or C_S.

Dealing first with the bipolar transistor, the stages in the analysis of the emitter decoupling capacitor circuit are shown in Fig. 6.15. Here the coupling capacitor C_{c1} is assumed to be effective at the frequency where C_E begins to have an effect on response. The impedance Z_E in the emitter circuit is R_E in parallel with $1/j\omega C_E$. Taking h_{oe} to be negligible, then the whole of $h_{fe}i_1$ (as well as i_1 itself) flows into Z_E and the voltage drop across Z_E appears in the 'input' loop; this voltage acts to oppose the signal input voltage. It has no effect, however, on the output circuit because of the constant current generator so the circuit reduces to that shown in Fig. 6.15(b). We can deduce from this that the current gain of the amplifier is not affected by Z_E but that the voltage gain and the input impedance are affected. The gain will begin to decline when the new term in the input impedance, $Z_E(h_{fe} + 1)$ becomes comparable with the rest of the input resistance, $(R_S + h_{ie})$. Applying Kirchhoff to the input loop and solving for current i_1, we have

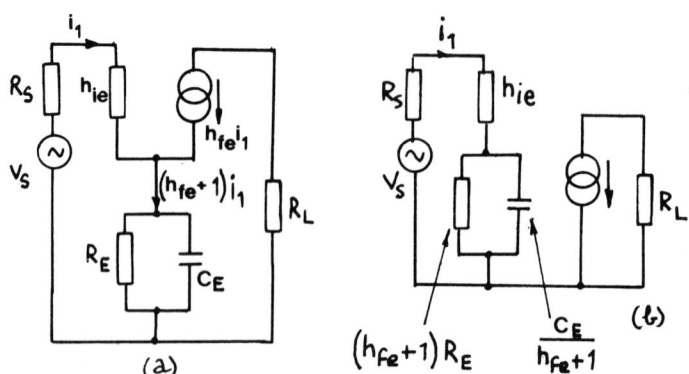

Figure 6.15

$$i_1 = \frac{V_s}{R_s + h_{ie} + Z_E(h_{fe} + 1)}$$

where

$$Z_E = \frac{h_{fe} + 1}{1/R_E + j\omega C_E}$$

$$= \frac{1}{\dfrac{1}{R_E(h_{fe} + 1)} + j\omega \cdot \dfrac{C_E}{h_{fe} + 1}}$$

To simplify things a bit we need to sort out a few typical figures: we can clearly replace $(h_{he} + 1)$ by h_{fe} for a beginning, and $R_E h_{fe}$ is going to be considerably greater than the total input circuit resistance $(R_s + h_{ie})$. So if $Z_E h_{fe}$ is to be small, then the low reactance of C_E must do the reducing - R_E cannot be reduced. Hence to a good approximation Z_E can be replaced by $1/j\omega C_E$ and the current through the effective large resistance $R_E h_{fe}$ ignored. Then

$$i_1 \simeq \frac{V_s}{R_s + h_{ie} - j\omega \dfrac{h_{fe}}{C_E}}$$

Current i_1 will drop to 0.707 of its mid-frequency value and the gain will drop to 0.707 A_o at a frequency ω_1 when the real and imaginary terms are equal:
Therefore

$$R_s + h_{ie} = \frac{h_{fe}}{\omega_1 C_E}$$

whence

$$\omega_1 = \frac{h_{fe}}{C_E(R_s + h_{ie})} \tag{6.7}$$

C_E is usually an electrolytic capacitor and is selected to provide the required low frequency response. If the coupling capacitors are then selected to be several times the values calculated for them from the earlier equations, the lower half-power frequency is determined by C_E only.

Example 6.2   In the circuit of Fig. 6.16, h_{ie} = 1.5 kΩ, h_{fe} = 100, and R_1 and R_2 may be considered to have negligible loading on the input circuit.   Obtain a figure for the lower cut-off frequency of this circuit and specify suitable values for the coupling capacitors C_1 and C_2.

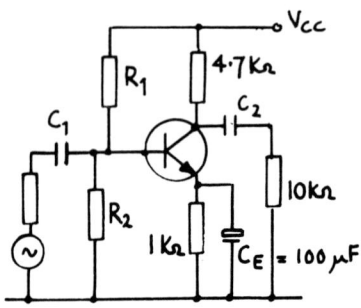

Figure 6.16

From equation (6.7)

$$\omega_1 = \frac{h_{fe}}{C_E(R_s + h_{ie})} = \frac{100}{100 \times 10^{-6} \times 3000}$$

$$= 334 \text{ rad/s}$$

Therefore

$$f_1 = \frac{334}{2\pi} = 53 \text{ Hz}$$

For this value of f_1 we find

$$C_1 = \frac{1}{\omega_1(R_s + h_{ie})} = \frac{10^6}{334 \times 3000} \text{ μF}$$

$$\simeq 1 \text{ μF}$$

A 10μ F capacitor could be used here.

$$C_2 = \frac{1}{\omega_1(R_L + R_o)} = \frac{10^6}{334 \times 14.7 \times 10^3}$$

$$\simeq 0.2 \text{ μF}$$

192

A 1 μF capacitor could be used here.

Example 6.3   A transistor with h_{ie} = 1.5 kΩ, h_{fe} = 66 and h_{oe} = 33 μS is used in common-emitter configuration with R_E first bypassed and then unbypassed. For R_E = 1 kΩ, R_L = 10 kΩ and R_S = 50 Ω, pre-dict A_i, A_v and R_i for the two cases.

The equivalent circuit is shown in Fig. 6.17 where this time h_{oe} is included. We will solve this problem by setting up loop equations and obtaining the current and voltage gains in terms of a general R_E. Working in kΩ and mA to avoid powers of 10, we have

$$v_s = i_1(1.55 + R_E) + i_3R_E \tag{i}$$

$$i_2 = 66i_1 - i_3 \tag{ii}$$

$$0 = R_E(i_1 + i_3) + 10i_3 - \frac{66i_1 - i_3}{33} \times 10^3 \tag{iii}$$

From the third of these we obtain (since $1/h_{oe}$ = 30 kΩ)

$$i_1(R_E - 2000) + i_3(R_E + 40) = 0$$

But R_E = 0 or 1 kΩ. For R_E = 0 (perfectly bypassed)

$$A_i = \frac{i_3}{i_1} = \frac{2000 - R_E}{40 + R_E} = \frac{2000}{40} = 50$$

For R_E = 1 (unbypassed)

$$A_i = \frac{1999}{41} = 48.76$$

The effect on the current gain is therefore negligible as we deduced a little earlier on.

For the voltage gain, taking A_i = 50, $i_1 = i_3/50$ and from equa-tion (i) above (taking $v_1 \simeq v_s$)

$$v_I = \frac{i_3}{50}(1.55 + R_E) + i_3R_E$$

$$= i_3 \cdot \frac{1.55 + R_E}{50} + R_E$$

But

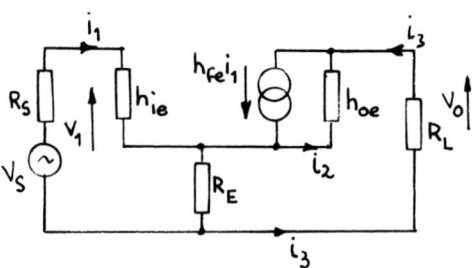

Figure 6.17

$$v_0 = -i_3 R_L = -10i_3$$

$$A_v = \frac{v_o}{v_i} = - \frac{50 \times 10}{1.55 + 51R_E}$$

For $R_E = 0$

$$A_v = - \frac{500}{1.55} = -322.6$$

For $R_E = 1$

$$A_v = - \frac{500}{52.55} = -9.5$$

The effect on the voltage gain is considerable. This is characteristic of negative feedback.

For the input resistance, we have

$$i_3 = 50i_1$$

and substituting this into equation (i)

$$v_1 = i_1(1.55 + R_E) + 50R_E i_1$$

$$R_i = \frac{v_1}{i_1} = (1.55 + R_E) + 50R_E$$

$$= 1.55 + 51R_E$$

For $R_E = 0$ $R_i = 1.55 \ k\Omega$

For $R_E = 1$ $R_i = 52.55$ kΩ

These values include the source resistance of 50 Ω. The effect of the feedback voltage across R_E on the input resistance is obvious.

From a study of the resulting expression for the input resistance with R_E unbypassed, we can surmise that R_i is approximately given by $(h_{ie} + h_{fe}R_E)$, so that, compared with the perfectly by-passed case where $R_i \simeq h_{ie}$, the input resistance is increased by the term $h_{fe}R_E$.

Since $h_{fe}R_E \gg h_{ie}$, $R_i \simeq h_{fe}R_E$. The voltage gain may therefore be expressed as

$$A_v = -h_{fe}\frac{R_L}{R_i} \simeq -h_{fe}\frac{R_L}{h_{fe}R_E} \simeq -\frac{R_L}{R_E} \tag{6.8}$$

This is an interesting result because (with the approximations in mind) we have an amplifier in which the voltage gain is independent of the parameters of the transistor.

We shall return to this topic in greater detail in Chapter 7.

Turning now to the FET, the source capacitor C_s is the important element in determining the lower cut-off frequency. We assume again that the coupling capacitors are effective at the frequency where C_s begins to have an effect on the response. The equivalent circuit for analysis is given in Fig. 6.18. Taking r_d to be very large compared with R_L, the whole of $g_m v_i$ flows into Z_E and develops a voltage that opposes, as in the case of the bipolar transistor, the signal voltage V_s. Working around the input loop we have

$$V_s - v_i = g_m v_i Z_s$$

so that

$$v_i = \frac{V_s}{g_m Z_s + 1} \tag{6.9}$$

and the voltage gain is reduced by the factor $1/(g_m Z_s + 1)$. When the gain is reduced to $0.707\, A_o$, $1 + g_m Z_s = \sqrt{2}$ and this is the lower cut-off frequency.

Let the reactance of C_s at this frequency be equal to the

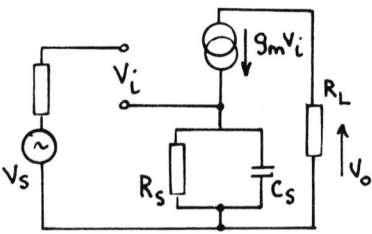

Figure 6.18

effective value of R_s - which is NOT just R_s, but the resistance seen looking at R_s with C_s connected across it. Putting $Z_s = R_s$, the open-circuit voltage across $R_s = g_m R_s v_i$ and substituting for v_i from (6.9) above yields

$$R_s = g_m R_s \frac{V_s}{g_m R_s + 1}$$

Since the current with R_s short-circuited is $g_m V_s$, we obtain the Thevenin equivalent as

$$\frac{V_{oc}}{I_{sc}} = \frac{R_s}{g_m R_s + 1} = \frac{1}{g_m + 1/R_s}$$

Hence the effective resistance in the source circuit is R_s in parallel with $1/g_m$. For a lower cut-off frequency ω_1, we then have

$$\omega_1 = \frac{1}{C_s (R_s \| 1/g_m)} = \frac{g_m + 1/R_s}{C_s} \tag{6.10}$$

Example 6.4   In the circuit of Fig. 6.19 the FET has g_m = 2.5 mS and the source resistance R_s is 2.7 kΩ. Find a suitable value for the bypass capacitor C_s if the lower cut-off frequency is to be 25 Hz.

Using (6.10) we have

$$C_s = \frac{(2.5 \times 10^{-3}) + (3.7 \times 10^{-4})}{2\pi \times 25} \times 10^{-6} \ \mu F$$

$$\simeq 18.3 \ \mu F$$

We can check on the lower cut-off frequency by using this value of C_s and f_1 = 25 Hz.

$$g_m Z_s = \frac{g_m}{1/R_x + j\omega C_s} = \frac{0.0025}{0.00037 + j0.0028}$$

$$= 0.116 - j0.87 = 0.88\underline{/-82.4^o}$$

Then

$$g_m Z_s + 1 = 1.116 - j0.87$$

$$1.41\underline{/-38^o}$$

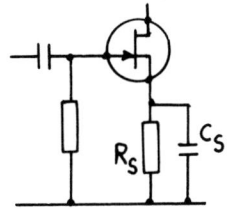

Figure 6.19

The magnitude of $g_m Z_s + 1$ is $1.41 \approx \sqrt{2}$, therefore f_1 = 25 Hz is the lower cut-off frequency with the 18.3 μF capacitor in circuit. In practice a 20 μF capacitor could be used, such an approximation being justified because of the wide tolerances associated with electrolytic capacitors.

6.7   BODE PLOT APPROXIMATION

Bode plots are often drawn to a fair degree of accuracy in laboratory experiments, but in everyday practice such precision is rarely warranted. Since the gain rolls off to an asymptote of gradient 6 dB/octave for a first order frequency response, it is a frequent practice to approximate the curve by a series of straight lines. This is particularly useful when a multistage amplifier response is being investigated, and different roll off rates are involved.

Essentially, the broken lines shown on the curves of Figs. 6.8 and 6.11 are such approximations; when a response curve is drawn in this way, the 3 dB frequency is located at the point where the line

197

goes round the 'corner' as it were, and the name 'corner' or 'break frequency' is applied to these points. Where the curve crosses the frequency axis and the gain is unity (0 dB), the frequency is called the 'crossing' or 'transition frequency' f_c. Fig. 6.20 shows a typical high frequency roll off approximated in this way.

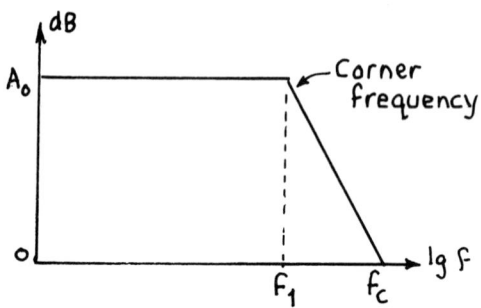

Figure 6.20

Example 6.5   An amplifier having a first-order frequency response has a mid-frequency voltage gain of 100 and an upper corner fre-quency of 25 kHz. Sketch an approximate Bode diagram of this ampli-fier. What is its voltage gain at 250 kHz?

The decibel gain = 20 lg 100 = 20 × 2 = 40 dB. At the transition frequency f_c, the gain is 0 dB. Therefore

$$20 \log A_0 = 20 \lg \frac{f_c}{f_2}$$

where f_2 is the 3 dB or corner frequency. Therefore

$$f_c = A_0 f_2$$

$$= 100 \times 25 \text{ kHz} = 2.5 \text{ MHz}$$

At a frequency of 250 kHz which is 1 decade above the corner frequency, the gain will be 20 dB down. Therefore

gain at 250 kHz = 40 - 20 = 20 dB.

The curve is sketched in Fig. 6.21.

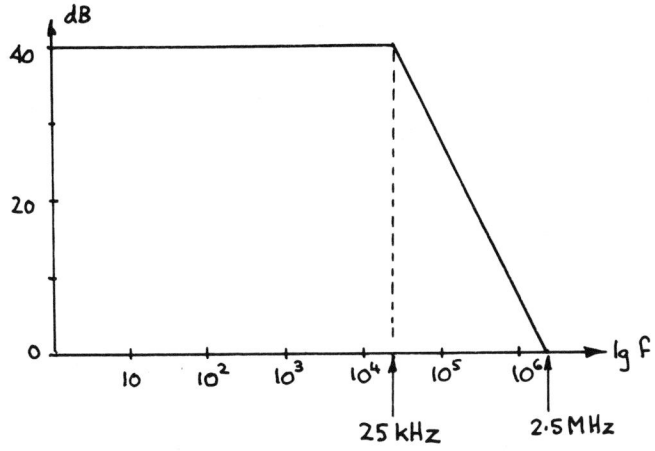

Figure 6.21

6.8   CASCADED AMPLIFIERS

Only rarely does a single stage provide the amplification required of
a system; two or more transistors are then 'cascaded', the output of
each stage providing the input to the next. Some problems arise
immediately, for not only must the input and output impedances of
the whole amplifier be of a suitable form to satisfy the requirements of the input and output loading, but the couplings from each
stage to the next must be properly established if the overall amplification is to be achieved.

The problem is simplified if the load of one stage is similar
to the input resistance of the next, that is, if $R_L = R_i$ (or h_{ie}).
This is roughly possible with bipolar transistors but not with FET's.
transformer coupling excluded. For bipolar transistors it follows
that since $A_v = -A_i R_L/h_{ie}$ then $A_v = -A_i$ and so the power gain can be
expressed as either A_v^2 or A_i^2. Clearly, no power gain is possible
if A_v or A_i is less than unity. This condition arises in common-
base mode, where $A_i < 1$ and in common-collector mode, where $A_v < 1$.

For a chain of connected amplifiers as shown in Fig. 6.22, the
overall voltage gain is

$$A = \frac{v_o}{v_i} = \frac{v_1}{v_i} \times \frac{v_2}{v_1} \times \frac{v_o}{v_2}$$

199

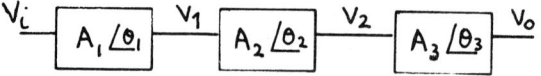

Figure 6.22

$$= A_1 \times A_2 \times A_3$$

or $\quad A \underline{/\theta} = A_1 A_2 A_3 \underline{/\theta_1 + \theta_2 + \theta_3}$

where both A and θ are functions of frequency. The overall current gain can be described in the same way.

The calculation of overall gain in cascaded amplifiers can be carried out using the usual h-parameter equations. Consider the two-stage equivalent circuit shown in Fig. 6.23. The parameters here are simply indicated h_i, h_f and h_o without designation of the configuration used, and the primes attached to these parameters indicate whether the first (single primes) or the second stage (double primes) is involved. If a load is connected to the output terminals and a source to the input terminals, the usual equations can be set up from which A_i, A_v, R_i and R_o can be established. The fact that different configurations may be employed for the stages is immaterial, the appropriate parameters for the configuration concerned being used as necessary. There are nine possible ways of connecting the three configurations as a pair; one set of equations will suffice for all of them.

If the Bode responses of the individual stages of the amplifier are known, a composite response can be easily derived. In this process the gain is best expressed in dB, for the overall response curve is then simply the ordinate sums of the individual response

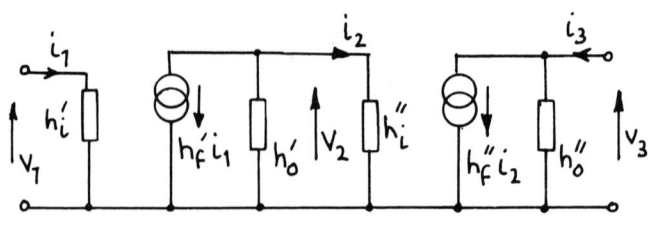

Figure 6.23

200

curves. Fig. 6.24 is an illustration. Here two stages are assumed to have the Bode plots marked as A_1 and A_2 and straight line approximations have been used. Amplifier A_1 has a mid-frequency gain of 40 dB and corner frequencies at 100 Hz and 100 kHz; amplifier A_2 has a mid-frequency gain of 30 dB and an upper corner frequency of 1 MHz. The low frequency response here extends down to d.c. The overall gain of the two stages when cascaded is the sum of A_1 and A_2 and is shown on the diagram as $A_1 + A_2$.

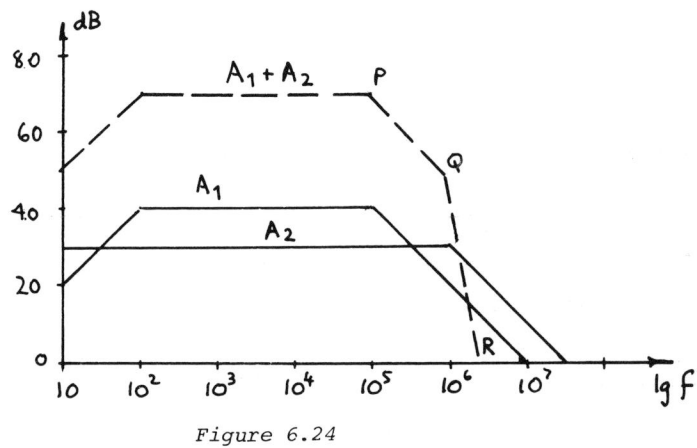

Figure 6.24

The point of interest in this diagram is the appearance of two break frequencies in the high frequency response roll off. The gradient of the response from P to Q is clearly 6 dB/octave, the same as that of amplifier A_1 over that part of the range; the gradient of the response from Q to R is 12 dB/octave, a combination of the 6 dB gradients of both A_1 and A_2. The addition of a further stage might well introduce another break frequency into the overall characteristic and this could lead to a section having a gradient of 18 dB/octave. Such gradients are characteristic of second and third-order filter circuits; in these the phase shift is not restricted to the 0°-90° that it is in the simple first-order system we have so far discussed in our study of couplings. Those regions in which the roll off is greater than 6 dB/octave have great significance when negative feedback is applied to a multistage amplifier, and this will be discussed later in the book.

6.8.1   Overall corner frequencies

The frequency response of a multistage amplifier may be a lot poorer than that of any of the individual stages making it up. A chain is no stronger than its weakest link and the response of a complete (untuned) amplifier is no wider (without feedback) than its narrowest part. Tuned amplifiers can be 'staggered' so that a response wider than any of the individual stages is obtained. In untuned amplifiers with roughly identical stages, the overall performance may be quite un-satisfactory even if the individual responses are adequate.

We have seen that the gain of an individual stage declines ideally with increasing frequency according to equation (6.6) given earlier:

$$A_H = \frac{A_o}{\sqrt{1 + [\frac{\omega}{\omega_2}]^2}}$$

where ω_2 is the upper 3 dB frequency. Now if we are going to have a cascaded amplifier with an overall gain A_T made up of n identical stages equated to the composite stages, then $A_T = (A_o)^n$ and

$$\frac{A_T}{\sqrt{2}} = \frac{(A_o)^n}{\sqrt{2}} = \left\{ \frac{A_o}{\sqrt{1 + [\frac{\omega}{\omega_2}]^2}} \right\}^n$$

From this

$$\sqrt[n]{2} = 1 + [\frac{\omega}{\omega_2}]^2$$

and then

$$\omega = \omega_2 [2^{1/n} - 1]^{\frac{1}{2}} \qquad (6.11)$$

This gives us the upper 3 dB frequency for n identical stages. The equivalent expression for the lower 3 dB frequency is easily derived (try it) and gives us

$$\omega = \frac{\omega_1}{[2^{1/n} - 1]^{\frac{1}{2}}} \qquad (6.12)$$

Example 6.6   Three stages in cascade each have an upper corner frequency of 25 kHz. What will be the upper corner frequency of the amplifier overall?

From equation (6.9)

$$= 25\sqrt{2^{1/3} - 1} = 25\sqrt{0.26}$$

$$= 12.75 \text{ kHz}.$$

The bandwidth is approximately halved relative to the individual bandwidths.

Example 6.7   For the amplifier shown in Fig. 6.25 working on a frequency where the effect of the coupling capacitors and bias resistors is negligible, calculate A_v and A_i, given that the transistors are identical and have the following h-parameters: $h_{ie} = 2 \text{ k}\Omega$, $h_{fe} = 100$, $h_{oe} = 33 \text{ }\mu\text{S}$, $h_{re} = 5 \times 10^{-4}$.

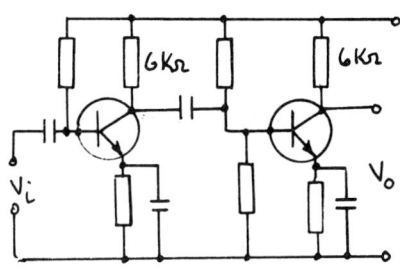

Figure 6.25

In many problems of this sort, h_{re} and h_{oe} can be neglected and the calculation is considerably eased without any serious effects. We include them here as an illustration.

The equivalent circuit is shown in Fig. 6.26(a) with its simplification at (b). Here parallel resistors have been combined into a single effective load on each stage. We can set up three defining equations from the first two sections of the circuit and one from the third:

$$v_1 = 2i_1 + (5 \times 10^{-4})v_2 \tag{i}$$

$$v_2 = -(i_2 + 100i_1)5 \tag{ii}$$

$$v_2 = (5 \times 10^{-4})v_o + 2i_2 \tag{iii}$$

$$i_2 = -v_o/500 \tag{iv}$$

Substituting for i_2 in (ii) and (iii) yields

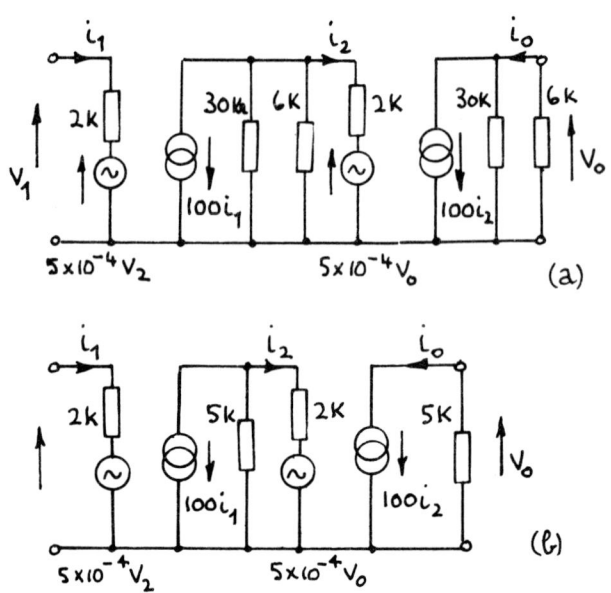

Figure 6.26

$$v_2 = -\left[\frac{-v_o}{500} + 100i_1\right] 5 = (100 \times 10^{-4})v_o + 2i_2 \tag{v}$$

Also

$$-\left[\frac{-v_o}{500} + 100i_1\right] 5 = (5 \times 10^{-4})v_o - \frac{2v_o}{500}$$

Therefore

$$(100 \times 10^{-4})v_o - 500i_1 = (5 \times 10^{-4})v_o - \frac{2v_o}{500}$$

$$= (5 \times 10^{-4})v_o - (40 \times 10^{-4})v_o$$

$$i_1 = \frac{135 \times 10^{-4}}{500} v_o = 2.7 \times 10^{-5}v_o \tag{vi}$$

Substituting for i_1 in (v)

$$v_2 = (100 \times 10^{-4})v_o - (135 \times 10^{-4})v_o$$

$$= -(35 \times 10^{-4})v_o \tag{vii}$$

Substitute from (vi) and (vii) into (i) and we have

$$v_1 = (5.4 \times 10^{-5})v_o - 5 \times 10^{-4}(35 \times 10^{-4})v_o$$

$$= (5.4 \times 10^{-5})v_o - (175 \times 10^{-8})v_o$$

Then

$$A_v = \frac{v_o}{v_1} = \frac{1}{5.2 \times 10^{-5}} = 19230 \ (86 \ dB)$$

Also from the equivalent circuit

$$v_o = 6i_o$$

and dividing this by (vi)

$$\frac{6i_o}{i_1} = \frac{v_o}{(2.7 \times 10^{-5})v_o}$$

Therefore

$$A_i = \frac{i_o}{i_1} = \frac{10}{6 \times 2.7} = 6173$$

6.9   A LABORATORY EXPERIMENT

This experiment illustrates the Miller Effect. Make up the circuit of Fig. 6.27 using a 2N3819 FET as a common-source amplifier. This amplifier will have a gain of about 15 to 20.

Capacitor C is a standard calibrated variable which should have a maximum capacity of the order of 1000 pF. Inductor L is chosen to resonate with C in a series circuit at about 5 kHz (with C set to about 1000 pF). The signal generator should have a low impedance output, preferably less than 50Ω and R_s is a resistor of this value.

With the amplifier attached and C set to about its maximum value, tune the signal generator to obtain resonance indicated by a maximum reading on the a.c. (high impedance) voltmeter V_i. Now connect the amplifier and adjust the signal generator output amplitude so that an undistorted output is obtained at the amplifier output. Retune

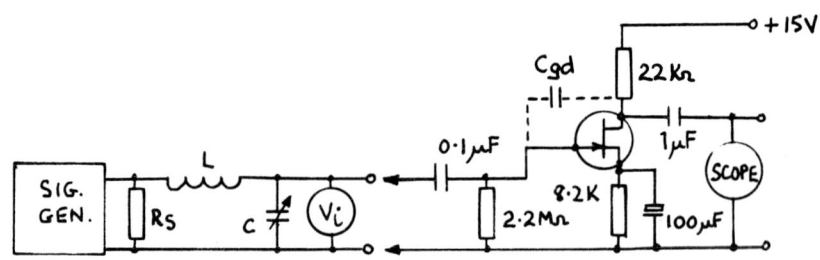

Figure 6.27

capacitor C to restore resonance and note the new capacitor reading.
This will be lower than the original reading, the difference giving
the input capacitance of the amplifier. Record this.

Now simulate various values of C_{gd} for the FET by connecting in
turn capacitors of 5 pF, 10 pF and 20 pF between the drain and gate
terminals of the FET. For each case, restore resonance by adjustment of C and note the effective input capacitance of the amplifier.

Determine the basic amplifier gain by any suitable method, let
this be A_v. From the Miller equation

$$C_i = C_{gd}(1 + A_v) + C_{gs}$$

verify, for each 'artificial' value of C_{gd} that the equation leads to
the measured value of C_i. Account for any discrepancies. What is
the actual value of C_{gd} for this transistor?

PROBLEMS 6

1. Distinguish between 'large' and 'small' signals. What does the
dividing line between them in a given circumstance depend on?

2. What difference is there, if any, between the mid-frequency range
and the bandwidth of an amplifier?

3. What do you understand by the 3 dB points on a gain-frequency
response curve? These points are commonly used to define bandwidth.
Comment on the merits or otherwise of this definition.

4. Distinguish between coupling capacitors and bypass (or decoupling)

capacitors.

5. What limits the low frequency response of a bipolar or FE transistor when used as an RC coupled amplifier?

6. What limits the high frequency response?

7. Briefly differentiate between a Bode response plot and a Nyquist plot. Comment on any advantages or disadvantages that either has relative to the other.

8. Find the frequency at which a parallel arrangement of 10 kΩ and 500 pF gives an 'output' that is 3 dB below its 'mid-frequency' value.

9. What is meant by a 'cascaded' amplifier? Why is it generally necessary to cascade amplifiers?

10. Explain why a Bode plot of amplifier gain A where $A = 1/(1 + j\omega CR)$ above the upper 3 dB point ω_2, where $\omega_2 = 1/CR$, is a straight line of gradient -6 dB/octave.

11. In the circuit of Fig. 6.28, h_{ie} for the transistor is 1 kΩ. What value of C is required if the lower 3 dB point is to be 25 Hz?

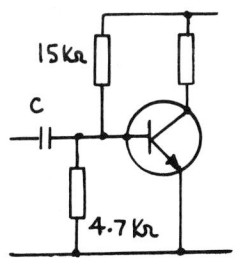

Figure 6.28

12. An amplifier has a response at 100 Hz which is 8.5 dB down on its response at 1 kHz. If the voltage gain at 100 Hz is 15, what is its voltage gain at 1 kHz?

13. The gain A of a single stage amplifier ideally declines with reference to the mid-frequency gain A_o as the frequency ω is

increased according to the relation

$$A = \frac{A_o}{1 + j\left(\frac{\omega}{\omega_2}\right)}$$

where ω_2 is the upper half-power frequency. Derive an expression for the upper half-power point of a multistage amplifier having N identical stages. Hence show that for three stages in cascade, each with f_2 = 20 kHz, the overall upper limit will be about 10.2 kHz.

14. An amplifier has a mid-frequency gain of 150 with upper and lower 3 dB points at 50 kHz and 15 Hz respectively. Sketch the Bode response curve for this amplifier, showing the gain and phase angle against frequency. What is the bandwidth of this amplifier?

15. A single stage FET amplifier must have a gain at 5 Hz equal to 90% of its mid-frequency gain. What lower half-power point is required?

16. A four stage amplifier is to have a bandwidth of 50 Hz to 15 kHz. Assuming identical stages, what must be the high and low cutoff frequencies of each stage?

7 Feedback

7.1 7.1   INTRODUCTION

The general idea of feedback has already been introduced; the base-
bias circuit of Fig. 3.11 and the effect of the unbypassed emitter
resistor of Fig. 6.17 are two examples. These have both been in-
stances of what is called 'local feedback' which pertains to feedback
that is applied to a single amplifying stage, and we shall return to
such circuits presently.

Feedback may take one of two forms: negative feedback (n.f.b.)
where the amplitude and phase of the signal fed back are such that
the overall gain of the amplifier is reduced; and positive feedback
where the amplitude and phase of the signal fed back are such that
the overall gain is increased. Unintentional feedback resulting
from poor circuit design is always positive feedback and the amplifier becomes quite unsuitable for its intended function in life.
Controlled feedback can be usefully applied only to an amplifier
which has been soundly designed in the first place. In this chapter
we shall be concerned with the theory and applications of negative
feedback.

Although a reduction in gain may not at first sound particularly
attractive, the employment of negative feedback gives a number of
significant improvements in amplifier performance and the method is
almost universally employed. It is a relatively simple matter to
make up for a loss in gain and the advantages are then a free bonus,
as it were, for the designer and user.

Among the more significant advantages in amplifier performance
resulting from negative feedback are:

1.   The overall gain can be made insensitive to device parameters, temperature variations, supply variations and the ageing of components.

2.   The gain can be stabilised over a wide frequency range.

3.   Noise, distortion and system disturbances can be significantly reduced.

4.   Input and output impedances can be modified and tailored to a particular design requirement.

7.2   THE FEEDBACK PRINCIPLE

Feedback is involved when ALL or a fraction of the output is fed back to the input of an amplifier. Here the word 'output' does not necessarily refer to the final stage output of the amplifier but may be the output from some intermediate point within the amplifier at which the feedback signal is derived. An amplifier (assumed to be stable) working without feedback has an 'open-loop gain' A_v; when feedback is operative, the overall or 'closed-loop' gain is A_v'. Other symbolisms are commonly used but we will retain these in this book. The fraction of the output amplitude fed back, be it current or voltage is designated β, and β is known as the feedback factor. Clearly, this will lie between 0 and 1. Do not confuse β as used here with the common-emitter short-circuit current gain discussed in an earlier chapter.

The block system of Fig. 7.1 will illustrate the elementary principle of feedback. Fig. 7.1(a) shows the amplifier without feedback applied; the open-loop amplifier forward gain is $A_v = v_o/v_i$. A_v may be positive or negative and may also be complex.

Fig. 7.1(b) shows the same amplifier with feedback applied. Here the feedback network connects together the input and output terminals of the amplifier and its output voltage $V_F = \beta v_o$ is applied in series with the input terminals of the amplifier. The β network will normally be positive but may also be complex. When the network is connected, the summation at the input terminals gives

$$v = v_i + \beta v_o$$

and the overall amplification of the system (or the external voltage gain) becomes

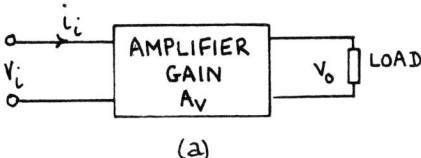

(a)

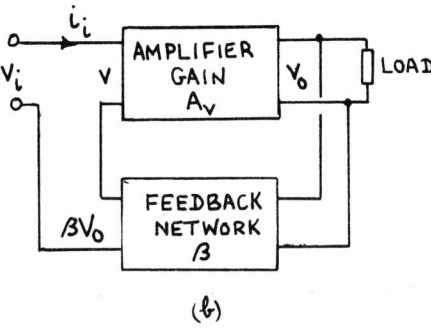

(b)

Figure 7.1

$$A'_V = \frac{v_o}{v_i} = \frac{v_o}{v - \beta v_o}$$

Dividing top and bottom by v yields

$$A'_V = \frac{A_V}{1 - \beta A_V} \tag{7.1}$$

This equation now needs interpreting in terms of the magnitude and sign of the denominator for the relationship between A_V and A'_V clearly depends upon this.

The quantity βA_V is the 'loop-gain' since it is the total gain measured around the feedback loop from input v to the output from the feedback network V_F, that is, from Fig. 7.1(b), $v_o = \beta A_V v_i$. In the absence of feedback, $\beta = 0$ and $A'_V = A_V$.

There are three possible conditions that the loop gain product may take which are of interest. If βA_V is positive and lies between 0 and 1 (but not equal to 1), the denominator of (7.1) will be less than unity and the overall amplification A'_V will be greater than A_V. This is positive feedback. If βA_V is negative, the denominator of

(7.1) will be greater than unity, the overall amplification will be less than A_v and the feedback will be negative or degenerative. If $\beta = 1$ so that $\beta A_v = A_v$, the entire output voltage is supplied to the input and A_v' must be less than unity.

A particular case arises when $\beta A_v = 1$. The resultant gain is infinite and the circuit has an output independent of any external input voltage. Strictly, equation (7.1) is invalid in this case since it was derived on the assumption that the amplifier was stable. Example 7.1   An amplifier has a gain of -10^4 and 1/100th of the output voltage is fed back in opposition to the input. Calculate the overall gain and the change in this gain if the basic gain of the amplifier dropped by 20%.

$$A_v' = \frac{-10^4}{1 - [-10^4/100]} \simeq -99$$

When the gain falls from 10^4 to -8×10^3

$$A_v' = \frac{-8 \times 10^3}{1 - [-\frac{8 \times 10^3}{100}]} \simeq -98.7$$

For a 20% variation in the forward gain, the overall gain of the feedback amplifier suffered only about a 0.3% change.

Notice from this example that the double negative that appears in the denominator identifies negative feedback by converting equation (7.1) to the form

$$A_v' = \frac{A_v}{1 + \beta A_v} \qquad\qquad (7.2)$$

7.3   EFFECTS OF NEGATIVE FEEDBACK

7.3.1   Gain stability

The previous example illustrated a very important aspect of negative feedback; the gain was held constant even though the forward gain of the basic amplifier altered very considerably. We can compare the performance of a feedback amplifier of gain A_v' with that of an unstabilised amplifier having an open-loop gain A_v; for the unstabilised amplifier, when a change in amplification occurs, clearly

$$\frac{dA_v}{dA_v} = 1$$

For the feedback amplifier, the expression for the fractional gain sensitivity can be obtained by differentiating (7.1):

$$\frac{dA_v'}{A_v} = \frac{1}{1 - \beta A_v} \cdot \frac{dA_v}{A_v}$$

We conclude from this that the feedback amplifier exhibits a much greater gain stability with respect to the internal factors that cause gain variations.

This aspect can also be looked at in a different way, for if $A_v \gg 1$, then

$$A_v' = \frac{A_v}{1 - \beta A_v} \simeq - \frac{A_v}{\beta A_v} = - \frac{1}{\beta}$$

This makes A_v' independent of A_v. Since β is normally fractional, the way to make βA_v large is to have A_v large. The requirement $\beta A_v \gg 1$ implies that the closed loop gain must be much smaller than the open-loop gain, hence heavy negative feedback leads to gain stability.

7.3.2   Harmonic distortion

There is no problem in obtaining distortionless amplification in small signal stages. In the power stages, however, the large signal excursions made necessary in the search for high efficiency usually introduce non-linear distortion. Such distortion arising WITHIN the amplifier can be reduced by the proper application of negative feedback.

Let an amplifier without feedback be considered as having harmonic distortion introduced in the form of a hypothetical generator connected as shown in Fig. 7.2(a). The distortion fraction is then D/v_o where D is the magnitude of the distortion voltage. Now let a feedback network be introduced as shown in Fig. 7.2(b). Since the distortion will in general be a function of the input voltage v, suppose v_i to be so adjusted that the input voltage at the amplifier

213

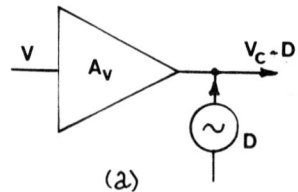

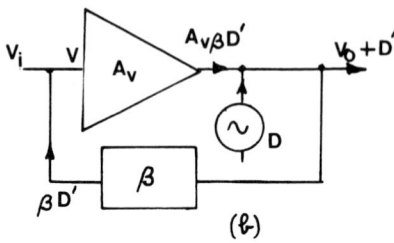

Figure 7.2

'terminals' is the same as before. Then the harmonic voltage fed back = $\beta D'$, say, and this appears at the output as $A_v \beta D'$. Then

$$D'(1 - \beta A_v) = D$$

since $\beta A_v D' + D = D'$. Hence $D'/D = 1/1 - \beta A_v$. So the distortion component is reduced by the factor of $(1 - \beta A_v)^{-1}$. In other words, provided that the output power is maintained at the same level, with and without feedback, then the introduction of negative feedback reduces the non-linear distortion component present at the output.

This argument may seem academic at first, for obviously the distortion is reduced by the same factor as is the gain of the system. It follows that if we reduce the amplifier gain by the application of feedback, the signal excursions will be proportionally reduced and the non-linearities will be less evident in the output. This is perfectly true, but the point being made is that signal reduction was not of concern in the analysis above. The distortion was introduced as a lumped term D, not as a function of the output amplitude. What is happening is that the distortion is being sampled and an amplified version of this sample is being fed back in antiphase to the original distorted waveform. Hence a portion of it is cancelled out. What must be kept in mind is that the output amplitude of the amplifier can be the same with or without feedback.

Example 7.2   A single RC coupled amplifier stage is connected in a circuit configuration as shown in Fig. 7.1. The open-loop gain of the amplifier is 10^3 and its distortion content is 5%. Estimate the distortion with negative feedback applied if the feedback ratio $\beta = 0.02$.

Evaluating $(1 + \beta A_v)$ gives us $1 + 10^3 \times 0.02 = 21$. Then

$$A'_v = \frac{10^3}{21} = 47.6.$$

The distortion with feedback is then $\frac{5\%}{21} = 0.24\%$.

7.3.3   Input impedance

An amplifier of open-loop gain A_v and input impedance Z_i is shown in Fig. 7.3(a). This impedance is the ratio of v_i to i_i. When feedback is applied in series with the input as shown in Fig. 7.3(b) and the feedback voltage βv_o is phased to oppose v_i, the effective voltage v at the amplifier terminals will be less than v_i and hence, as seen from these terminals, the input current will be reduced from i_i to some smaller value i'_i. This implies that the input impedance as seen from the 'source' terminals has been increased.

With the feedback loop closed we have $Z'_i = v_i/i'_i$ and $v_i + \beta v_o = i'_i Z'_i$.

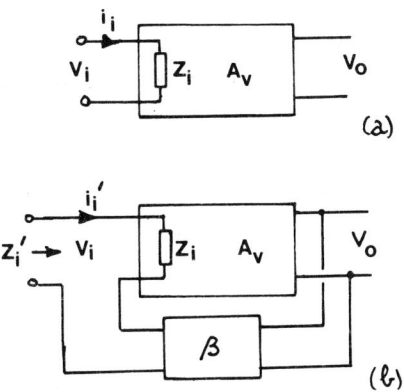

Figure 7.3

But

$$v_o = A_v' v_i = \frac{A_v v_i}{1 - \beta A_v}$$

Therefore

$$v_i \left[1 + \frac{A_v}{1 - \beta A_v} \right] = i_i' Z_i$$

Therefore

$$Z_i' = \frac{v_i}{i_i'} = Z_i (1 - \beta A_v) \tag{7.3}$$

The presence of the $(1 - \beta A_v)$ term indicates an increase in the input impedance due to the application of negative feedback, since the loop gain βA_v will be negative and $(1 - \beta A_v)$ will always be greater than unity.

7.3.4   Output impedance

Let the output circuit of the amplifier be represented as a voltage generator in series with its own internal (or output) impedance, Z_o. If the open- and short-circuit voltage and current at the output terminals are then measured or calculated, their ratio will give the impedance seen looking back into these terminals.

In Fig. 7.4(a) the open-circuit voltage $v_{oc} = e_o'$, where e_o' is the output e.m.f. with feedback applied. Hence $v_{oc} = A_v v_i / (1 - \beta A_v)$. In Fig. 7.4(b), on short-circuit there will be no feedback voltage so that $e_o' = e_o = A_v v_i$ and the short-circuit current will be $A_v v_i / Z_o$.

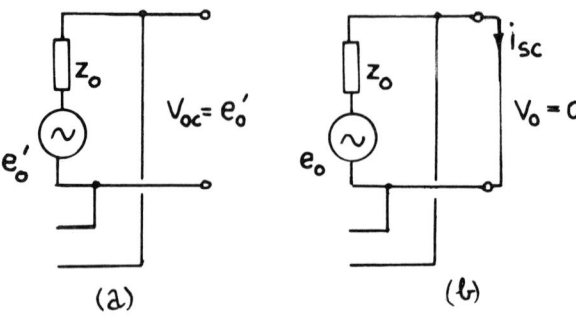

(a)   (b)

Figure 7.4

216

Hence

$$Z_o' = \frac{v_{oc}}{i_{sc}} = \frac{Z_o}{1 - \beta A_v} \tag{7.4}$$

We conclude that the output impedance is reduced by a factor $(1 - \beta A_v)^{-1}$ by the addition of negative feedback derived from the output voltage.

Example 7.3   An amplifier having Z_i = 1 kΩ and open-circuit output impedance 10 kΩ has a voltage gain of 800 when loaded with a 4 kΩ resistive load. A feedback voltage is derived from a potential divider in parallel with the load as shown in Fig. 7.5, where R_1 = 9.9 kΩ and R_2 = 100 Ω. Find the overall gain with feedback and the modified input and output impedances.

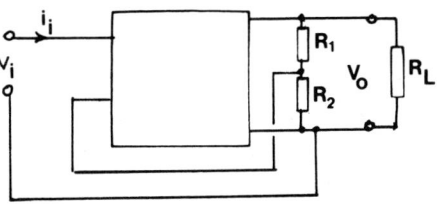

Figure 7.5

Without feedback

$$i_i = \frac{10^{-3}}{2 \times 10^3} \; A = 0.5 \; \mu A$$

$$v_i = i_i Z_i = 0.5 \times 10^{-6} \times 10^3 \; V$$

$$= 0.5 \; mV$$

Since A_v = 800,

$$v_o = 800 \times 0.5 \; mV = 0.4 \; V$$

$$i_o = \frac{0.4}{4 \times 10^3} = 100 \; \mu A$$

To determine the effects of the feedback, we first find the open-

217

circuit voltage gain, A_{oc}:

We have

$$800 = A_{oc} \frac{4}{4 + 10} = 2800$$

Now $R_1 + R_2 = 10$ kΩ and this is in shunt with the output load of 4 kΩ. Hence the effective load is 2.86 kΩ. Therefore the 'loaded' voltage gain is

$$\frac{2800 \times 2.86}{10 + 2.86} = 623$$

Also, since $R_2 \gg (Z_i + Z_s)$, the feedback fraction β is given by $R_2/R_1 + R_2 = 0.01$. Then

$$A_v' = \frac{623}{1 + 6.23} = 86.2$$

$$Z_i' = 1(1 + \{0.01 \times 623\}) = 7.23 \text{ k}\Omega$$

With the feedback applied, the new input current is

$$\frac{10^{-3}}{1 + 7.23} \text{ A} = 122 \text{ }\mu\text{A}$$

the overall input voltage $= 0.122 \times 7.23 = 0.89$ mV

$$v_o = 0.89 \times 86.2 = 76.7 \text{ mV}$$

and the new output current $= \dfrac{76.7}{4 \times 10^3}$ mA $= 19$ μA

Therefore

$$Z_o' = \frac{10}{1 + 6.23} = 1.38 \text{ k}\Omega$$

7.3.5   Bandwith

To investigate the effects of negative feedback upon the bandwidth of a circuit, we must assign frequency dependence to the forward gain. We have seen that

$$A_H = \frac{A_o}{1 + j\dfrac{\omega}{\omega_2}}$$

where A_H is the gain at high frequencies relative to the mid-frequency gain A_o and the 3 dB frequency ω_2. Substituting this into the feedback equation (7.1) yields

$$A_H' = \frac{A_o}{1 + j\omega/\omega_2 + \beta A_H}$$

and dividing through by $(1 + \beta A_o)$ we get

$$A_H' = \frac{A_o/(1 + \beta A_o)}{1 + j\{\omega/\omega_2(1 + \beta A_o)\}}$$

But $A_o/(1 + \beta A_o) = A_o'$. Hence

$$A_H' = \frac{A_o'}{1 + j\{\omega/\omega_2(1 + \beta A_o)\}} \tag{7.5}$$

We have used the positive sign here as we are clearly working with negative feedback.

Now the new upper 3 dB frequency is defined as the frequency at which $|A_H| = A_o'/\sqrt{2}$; let this frequency be ω_2', then from (7.5)

$$\omega_2' = \omega_2(1 + \beta A_o) \tag{7.6}$$

So the upper cut-off frequency has been increased by our familiar factor $(1 + \beta A)$, and the bandwidth correspondingly increased.

By using the relationship

$$A_L = \frac{A_o}{1 - j\dfrac{\omega_1}{\omega}}$$

where ω_1 is the lower 3 dB frequency, you should be able now to show that with negative feedback the new lower cut-off point ω_1' can be expressed by

$$\omega_1' = \omega_1/(1 + \beta A_o) \tag{7.7}$$

You will realise that what we have up to the present taken as A_V is, in general, the mid-frequency reference gain A_o. It might be instructive for you to turn back to Example 6.6 in the previous chapter and re-calculate the upper cut-off frequency by using equation (7.6). Can you account for the discrepancy in the answers?

7.3.6   Gain-bandwidth product

The form of the expression for the upper 3 dB point ω_2' with feedback applied given in (7.6) above, shows that a Bode plot derived from it will have the same 'shape' as for the case without feedback, but the mid-frequency level will be different and the 3 dB frequency will be higher, ω_2' instead of ω_2. Fig. 7.6 shows Bode plots of an amplifier with and without feedback.

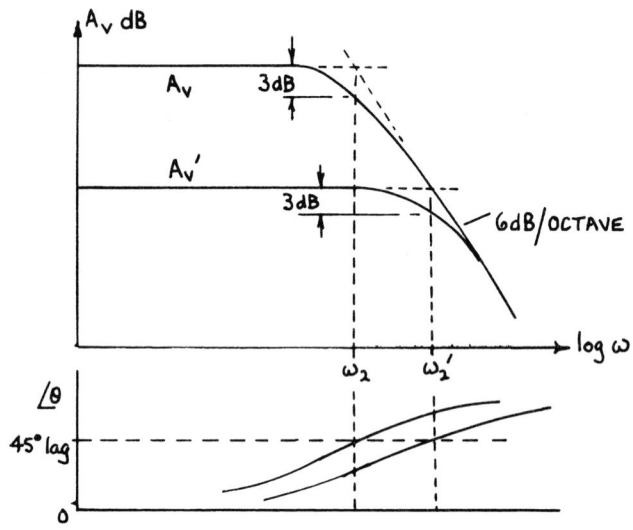

Figure 7.6

The point to notice is that the bandwidth increases from ω_2 to ω_2', a factor of $(1 - \beta A_o)$. This is exactly the factor by which A_o' is less than A_o. Thus the gain-bandwidth product for the amplifier remains constant. This is an important aspect of amplifier systems having a first-order roll off characteristic of 6 dB/octave since it enables the change in bandwidth to be assessed as feedback is applied. Trading gain for bandwidth is a very important practical concept. Any amount of gain (within reason) can be obtained from an amplifier by adding further stages, so the loss of gain through the application of feedback is not particularly significant. Restriction on bandwidth, however, is a fundamental physical property of an amplifier and negative feedback is a useful way of extending it to a desired value.

Although negative feedback will extend the bandwidth of amplifiers with higher order characteristics, these characteristics show a much more rapid fall in gain beyond the break frequency and the extension of bandwidth by the use of feedback is not so marked as it is in the first-order system. This is why a 6 dB/octave roll off is preferred. Another reason, and perhaps a stronger one, is that such an amplifier is and remains unconditionally stable when negative feedback is added to it; this aspect will be discussed presently.

7.4   CURRENT DERIVED FEEDBACK

In the previous discussion we have dealt only with voltage-derived feedback (or shunt feedback as it is sometimes called). The output voltage is sampled, operated upon by the feedback network, and connected back to the input circuit where it is added (algebraically) in series with the input signal.

It is possible to design current-derived feedback in which the output load current is sampled and fed back to the input by way of the feedback network, where it is combined with the input current from the signal source.

Example 7.3 above gave an illustration of how voltage derived feedback was obtained at the output; a simple potential divider made up of the resistors R_1 and R_2. By altering the ratio between these, any value of the feedback factor β can be obtained between zero and unity.

To feed back a signal proportional to the output current requires that a resistor is placed in series with the load as shown in Fig. 7.7(a). Assuming that $R_1 \ll R_L$, then the voltage fed back $V_F = i_o R_1$ and the introduction of R_1 has negligible effect on the output loading. The feedback fraction

$$= \frac{i_o R_1}{i_o(R_1 + R_L)} = \frac{R_1}{R_1 + R_L} \simeq \frac{R_1}{R_L}$$

At the input port, a current summation takes place, so a parallel connection is made. Current feedback lowers the input impedance and

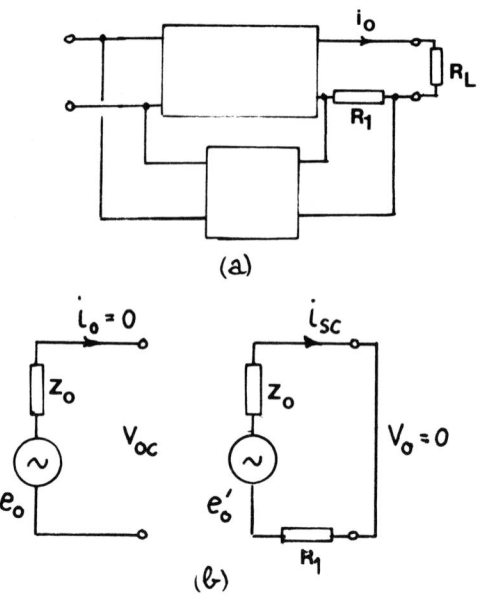

Figure 7.7

increases the output impedance of an amplifier system. Otherwise, its use is equivalent to the voltage-feedback previously discussed.

7.4.1   Effect on the impedances

Consider the output system of Fig. 7.7(a) under open- and short-circuit conditions; see Fig. 7.7(b). On open-circuit there is no feedback since $i_o = 0$, therefore $v_{oc} = e_o = A_v v_i$. On short-circuit, $R_1 \ll Z_o$, then

$$i_{sc} = \frac{e'_o}{Z_o} = \frac{A_v v_i}{1 + \beta A_v} \cdot \frac{1}{Z_o}$$

Then

$$z'_o = \frac{v_{oc}}{i_{sc}} = Z_o (1 + \beta A_v) \tag{7.8}$$

Hence the output impedance is increased. It is important to note that this increase has nothing to do with the simple addition of R_1 in series with the load. The same point applies in relation to

voltage feedback; the output impedance in that case is not decreased
because the feedback elements are connected in shunt with the output.

For the effect of parallel connected feedback at the input ter-
minals, we have from Fig. 7.8 $i_i = v_i/Z_i$. Then $Z_i' = v_i/i_i' = v_i/(i_i + i_f)$ where $i_f = V_f/Z_i$.

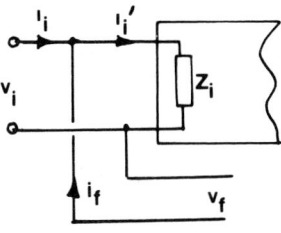

Figure 7.8

But $i_i = v_i/Z_i$ and so

$$Z_i' = \frac{v_i}{\dfrac{v_i}{Z_i} + \dfrac{v_f}{Z_i}} = \frac{Z_i v_i}{v_i + \beta A_v v_i}$$

Therefore

$$Z_i' = \frac{Z_i}{1 + \beta A_v}$$

Thus, with parallel connected feedback, the input impedance is
always reduced.

Example 7.4   Show that the application of negative feedback reduces
amplifier phase shift.

An amplifier has a voltage gain of 3.61 + j0.9 in the absence of
feedback. A non-reactive potential divider provides 44.6% of the output voltage to a feedback network that reduces its magnitude by 50%
and advances its phase by 60°. The resultant feedback voltage is
added to the input voltage and applied to the amplifier input.
Estimate the gain and phase angle of the amplifier with the feedback
and state whether this is an example of positive or negative feedback.

This example illustrates an important aspect of feedback problems.
We have discussed the basic expression for feedback, equation (7.1),
in the general sense that the denominator $(1 - \beta A_v)$ applies to

223

positive feedback and that, if the feedback is negative, A_V is negative and $A_V' = A_V/(1 + \beta A_V)$. This can be a risky assumption if βA_V has a complex value. It is best to use the basic equation in such cases and leave it to the mathematics to show whether the feedback is indeed positive or negative.

However, to the problem. Let the amplifier gain $A_V = |A_V| \underline{/\theta}$. Then with feedback

$$A_V' = |A_V'| \underline{/\theta'} = \frac{A_V}{1 + \beta A_V}$$

$$= \frac{A_V \underline{/\theta}}{1 + \beta |A_V| \underline{/\theta}}$$

$$= \frac{|A_V| \underline{/\theta}}{1 + \beta |A_V| \cos \theta + j\beta |A_V| \sin \theta}$$

Hence

$$\theta' = \theta - \tan^{-1} \frac{|A_V| \sin \theta}{1 + \beta |A_V| \cos \theta}$$

The application of negative feedback has reduced the phase shift by an angle

$$\tan^{-1} \frac{|A_V| \sin \theta}{1 + \beta |A_V| \cos \theta}$$

For the calculation:

$$A_V = 3.61 + j0.9 = 3.72 \underline{/14^o}$$

The feedback voltage $= 0.446 \ v_o$

But $\beta = 0.5 \underline{/60^o}$, therefore the voltage at the input terminals $= v_i + 0.223 v_o \underline{/60^o}$, since the 'overall' feedback voltage $= (0.5 \underline{/60^o})$ $(0.446 v_o)$.

Now

$$A_V' = \frac{3.72 \underline{/14^o}}{1 - (0.223 \underline{/60^o})(3.72 \underline{/14^o})}$$

$$= \frac{3.72 \underline{/14^o}}{1 - (0.83 \underline{/74^o})} = \frac{3.72 \underline{/14^o}}{1 - (0.23 + j0.8)}$$

224

$$= \frac{3.72\underline{/14^{\circ}}}{1.11\underline{/46^{\circ}}} = 3.35\underline{/-32^{\circ}}$$

Since both the gain and phase angle are reduced, the feedback is negative.

7.5   INSTABILITY IN FEEDBACK SYSTEMS

All the benefits of negative feedback are obtained at the expense of gain. The gain can be restored by the addition of further small signal stages but this may well introduce a drawback; the phase shift around the loop of a multi-stage amplifier may be such that the system will become unstable.

Over the mid-frequency range the phase shift remains constant, and provided the feedback voltage is arranged to be in phase opposition to the input signal there is normally no problem. At the low and also at the high frequencies, the forward and feedback elements become frequency sensitive, so that the output voltage is shifted in phase and changed in magnitude relative to the mid-frequency values. It is then possible for the feedback to become positive at certain low or high frequencies, resulting in regeneration and oscillation. This problem can only be eliminated by ensuring that the loop gain βA_v is less than unity at the frequencies concerned. As we have noted previously, when $\beta A \to 1$, $A_v' \to \infty$; this is a condition leading to an uncontrolled (oscillatory) output limited only by the cut-off and saturation limits of the amplifier devices.

Consider the high-frequency end of the response. From equation (6.6) earlier, we have for the high frequency gain A_H

$$A_H = \frac{A_o}{1 + j\omega/\omega_2}$$

At some frequency above ω_2, say, where $\omega = \sqrt{3}\omega_2$

$$A_H = \frac{A_o}{1 + j\sqrt{3}} = \frac{A_o}{2}\underline{/-\tan^{-1}\sqrt{3}} = \frac{A_o}{2}\underline{/60^{\circ}}$$

Each stage of an amplifier is now frequency sensitive, so for three stages in cascade the individual 60° shifts add up to 180° i.e.

225

$$A_H \text{ for three stages} = \left[\frac{A_o}{2}\right]^3 \underline{/-180^o} = -\left[\frac{A_o}{2}\right]^3 + j0$$

If we assume that $\beta = 0.01$ in this amplifier, then

$$\beta A_H = -0.01 \left[\frac{A_o}{2}\right]^3 + j0$$

$$= -\left[\frac{A_o}{9.3}\right]^3 + j0$$

For instability, A'_H must be infinite or $A_H = 1 + j0$. Therefore, $A_o = -9.3$.

So in a three stage amplifier, if the gain per stage is -9.3 and the feedback fraction is 0.01, the circuit will be unstable at a frequency for which the phase shift per stage is 60°.

7.5.1   The Nyquist plot

A convenient (though often tedious) way of testing the stability, or otherwise, of a feedback amplifier is to plot the loop gain βA_v on polar co-ordinates and examine the variation of the loop gain with frequency. Loop gain is a complex quantity and, as we have already seen, it can be represented in the manner of Fig. 7.9. Each complex phasor (or complexor as they are sometimes called in this context), gives the loop gain in magnitude and phase at a particular frequency. The line joining the ends of the complexors traces out the Nyquist locus and the complete diagram covers all frequencies from zero to

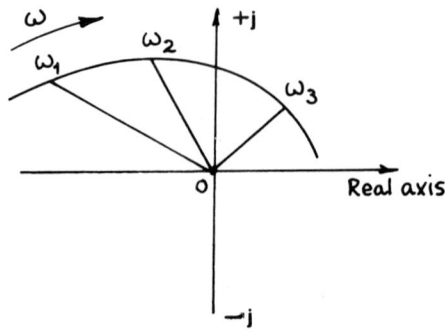

Figure 7.9

226

infinity. Nyquist's criterion (in a simplified form) gives this information: let a circle of unit radius and centre $(1 + j0)$ be drawn on polar axes, Fig. 7.10. If a point P on the polar plot of loop gain in the complex plane lies inside this circle, then the complexor length $(1 - \beta A_v)$ is <1, whereas if the point P lies outside the circle, the converse is true. Now $|A_v'| < A_v$ if $(1 - \beta A_v) > 1$ and the feedback is negative. This condition is satisfied for all βA_v for which the locus lies outside the unit circle. Inside the circle $|1 - \beta A_v| < 1$ and the gain is increased. Therefore the whole area within the unit circle corresponds to positive feedback. Hence, in the figure, the applied feedback is positive for the frequency range of ω_2 to infinity, while it is negative for the frequency range ω_1 to ω_2. The Bode plots of the frequency response of the gains with and without feedback can therefore be illustrated as in Fig. 7.11.

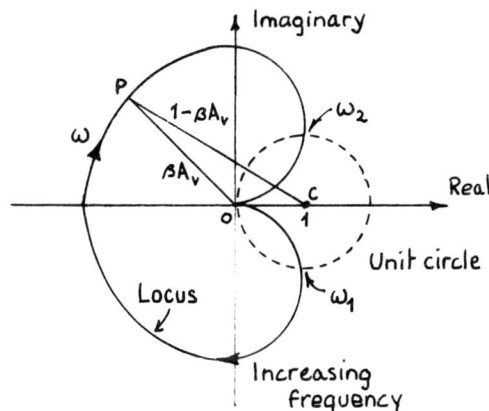

Figure 7.10

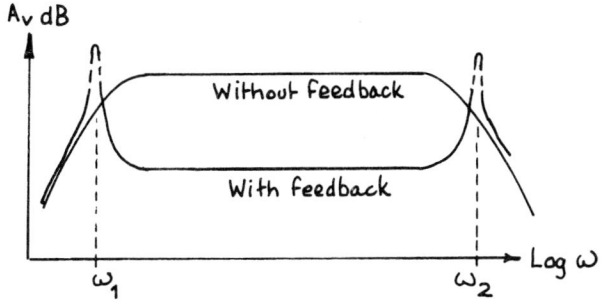

Figure 7.11

227

If $(1 - \beta A_v)$ is zero at a particular frequency, then PC is zero and the locus on which P lies will pass through C. Therefore, if the locus passes through C the circuit will become unstable. The point $(1 + j0)$ is referred to as the critical point and the frequency at which the locus passes through it is the oscillation frequency. As it is very unlikely that the locus would pass exactly through the critical point in a practical design, an extension of Nyquist's criterion for stability states that if the locus passes through C or 'encloses' C, the circuit will be unstable. In view of this we can deduce that the locus of Fig. 7.12(a) represents a stable feedback circuit while the locus of Fig. 7.12(b) represents an unstable circuit.

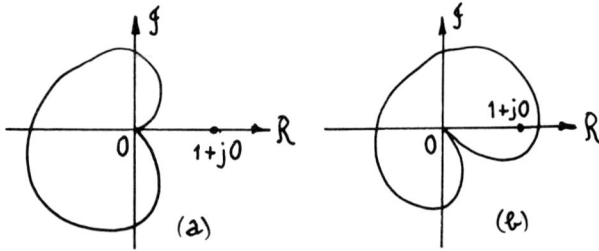

Figure 7.12

A special case arises when the Nyquist plot takes a form as shown in Fig. 7.13. Here the critical point is not cut or enclosed by the locus; hence the circuit appears (and is for this condition) stable. However, if the plot is caused to shrink uniformly at all frequencies by a reduction in gain occurring for same reason, a point

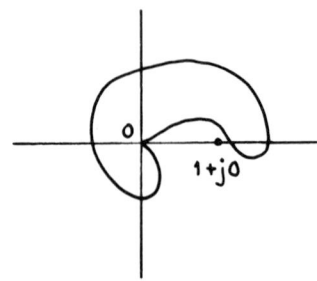

Figure 7.13

will be reached at which the critical point enters the loop and the circuit becomes unstable. If the plot shrinks still further, another stage is reached where the circuit becomes stable again. Circuits exhibiting such an effect are said to be 'conditionally stable'.

Conditional stability, which is often characteristic of amplifier using large amounts of feedback, is important for several reasons. If an excessive signal is momentarily applied to the system, the amplifier may saturate and its effective amplification fall. Under conditions of conditional stability, the momentary loss of gain may send the circuit into oscillation. Such oscillations may build up and the amplifier may remain saturated (and therefore unstable) even after the excessive input signal has gone. Also, when an amplifier is first switched on, the gain is zero. As the gain builds up, if the system is conditionally stable, it may have to pass through the unstable condition. In such a case oscillations will start and may grow to a sufficient magnitude to saturate the amplifier; a stable state is then never reached.

Example 7.5   The open-loop gain of an amplifier is $1000/80^\circ$ and $\beta = -0.01/10^\circ$. Calculate the feedback gain of the amplifier. What limiting value and phase of β would be required for the amplifier to become unstable?

$$\beta A_v' = \frac{1000/80^\circ}{1 + (1000/80^\circ)(0.01/10^\circ)}$$

$$\beta = \frac{1000/80^\circ}{1 + 10/90^\circ} = \frac{1000/80^\circ}{10.04/84.3^\circ}$$

$$= 99.5/-4.3^\circ$$

For instability

$$\beta A_v = 1/0^\circ$$

$$\beta = \frac{1/0^\circ}{1000/80^\circ}$$

$$= 0.001/-80^\circ$$

Example 7.6   An amplifier with $Z_o = 1$ kΩ has an overall voltage gain of $10^4/180^\circ$ when connected to a 1 kΩ load. The overall gain is

229

to be reduced to 200 by the simultaneous application of current and voltage feedback in series with the input so that Z_o is unaffected. Obtain a suitable value for the required current feedback resistor needed in series with the output load, and the percentage voltage feedback

$$A_V = 10^4/180^o = -10^4 + j0$$

$$A'_V = \frac{-10^4}{1 - \beta 10^4} = 200$$

From this we obtain $\beta = 0.0051$

Let the current feedback fraction be β_c and the voltage feedback fraction be β_V. Then the output impedance 'change' will be in the ratio

$$\frac{1 - \beta_c A_V}{1 - \beta_V A_V}$$

from equations (7.3) and (7.4) earlier. But if Z_o is not to change, $\beta_c = \beta_V = \beta/2 = 0.00255$, and the percentage voltage feedback is 0.225%.

Assuming the feedback resistor in series with the load is small relative to the load $\beta_c = R/1000$. Therefore $R \simeq .00255 \times 1000 = 2.55\Omega$.

7.6 SOME FEEDBACK CIRCUITS

We have already encountered circuit systems where operation depends upon negative feedback. The base bias arrangement shown in Fig. 3.11 was one example. By returning the base resistor R_B to the collector instead of the V_{CC} line as Fig. 7.14 indicates, any tendency for the collector current to increase lowers the static collector potential and the base current changes correspondingly. This in turn modifies the collector current in such a direction that the original change is nullified. In other words, an increase in I_c is fed back in such a way as to oppose the change that produced it. There is, of course, some negative feedback on the signal output with this system, and R_B is often split and decoupled with a suitable capacitor to avoid a.c. feedback.

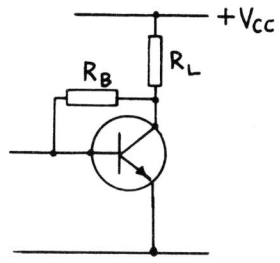

Figure 7.14

Another earlier example was illustrated by Fig. 6.17 and discussed in Example 6.3. The effect of leaving the emitter resistor
unbypassed led to the signal voltage developed across that resistor
to be subtracted from the input voltage, so reducing the stage gain
and substantially modifying the input resistance.

We shall now follow these examples up with a few common examples
of practical feedback systems: 'the followers'.

7.6.1   The source follower

In the circuit of Fig. 7.15, the load resistor R_L is taken from its
more familiar place in the collector or drain circuit of the amplifying device and placed in series with the emitter or source. As the
collector or drain is then returned to V_{CC} (an a.c. earth), the configuration is common-collector or common-drain. These circuits go
under the more common names of emitter-follower or source follower.
We consider first the FET circuit or the source follower.

Consider Fig. 7.16, where the FET has parameters g_m, r_d and μ and
the output signal is derived across the source load R_L. When v_i goes

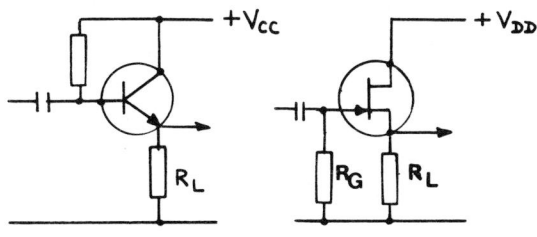

Figure 7.15

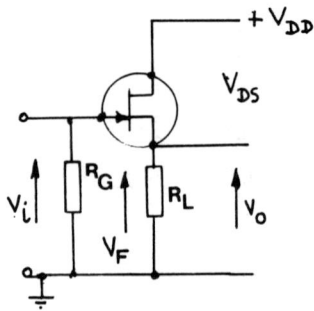

Figure 7.16

positive the source also goes positive, since i_D increases, hence v_o and v_i are in phase. However, in the gate circuit input relative to the 'source', v_i and the feedback voltage v_F developed across R_L are effectively in series; hence the gate-source potential is the difference between v_i and v_F and these voltages are therefore in phase opposition. Since $v_F = v_o$, this may appear contradictory, but in one case the gate is referenced to earth and in the other it is referenced to the source. So, for v_o in phase with v_i, the gain $A_v = v_o/v_i$ is positive; and for v_F antiphase to v_i and hence to v_o, the feedback factor $\beta = v_F/v_o$ is negative, i.e. $\beta = -1$.

This is an example of 100% negative feedback. For any amplifier we have

$$\delta i_D = g_m \delta v_i + \frac{1}{r_d} \delta v_{DS}$$

Hence, since

$$\delta v_i = \delta v_{DS} = -\delta v_F$$

$$i_D = g_m(-v_F) + \frac{1}{r_d}(-v_F)$$

Therefore

$$-\frac{\delta i_D}{\delta v_F} = g_m + \frac{1}{r_d}$$

The output impedance is consequently $\dfrac{1}{g_m + 1/r_d} \approx \dfrac{1}{g_m}$

This will clearly be small.

We now require to determine the voltage gain. From the equivalent circuit of Fig. 7.17 we have

$$i_D = \frac{\mu(v_i - v_F)}{r_d + R_L}$$

$$v_F = i_D R_L = \frac{\mu R_L (v_i - v_F)}{r_d + R_L}$$

Therefore

$$v_F(r_d + R_L) = \mu R_L(v_i - v_F)$$

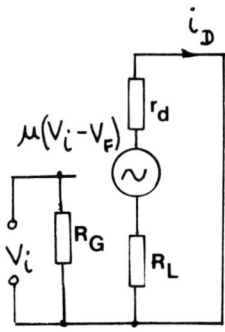

Figure 7.17

$$v_F(r_d + R_L + \mu R_L) = \mu R_L v_i$$

Therefore

$$A'_v = \frac{v_F}{v_i} = \frac{\mu R_L}{r_d + (1 + \mu)R_L}$$

$$= \frac{g_m R_L}{1 + g_m R_L} \tag{7.9}$$

since $\mu = r_d g_m$. This result must be less than unity.

The simple source follower of Fig. 7.16 has a disadvantage in that the output signal excursion is restricted. This comes about because of an inappropriate d.c. operating point. Under proper conditions the source should be at a potential midway between V_{DD} and earth, but with the gate returned to earth by way of R_G, the source cannot rise much above a volt before the transistor cuts off. A solution to this problem is to divide the load resistor into two parts, say, R_1 and R_2, and return R_G to the junction as shown in Fig. 7.18. This not only puts the d.c. conditions right but increases the already very high input impedance.

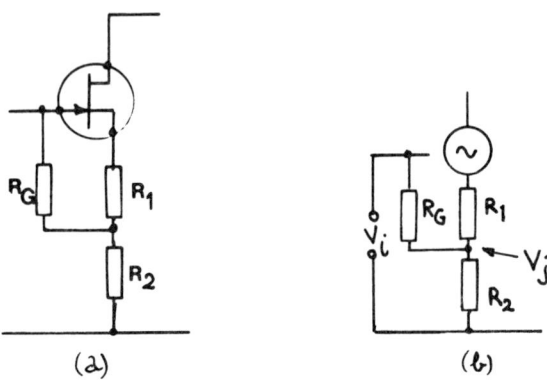

(a)   (b)

Figure 7.18

Looked at qualitatively, we can see that the lower end of R_G, being no longer tied to earth, now swings up and down in phase with the input signal. Thus R_G draws very little current from the input and its effective resistance appears correspondingly greater. Quanti-tively, referring to the equivalent circuit of Fig. 7.18(b) and neglecting the small current flowing in R_2 from the input, the poten-tial at the junction of the source resistors is

$$v_j = \frac{R_2}{R_1 + R_2} v_o = \frac{R_2}{R_1 + R_2} \cdot A_v \cdot v_i$$

so that

$$z_i = \frac{v_i}{i_i} = \frac{v_i}{(v_i - v_j)/R_G}$$

$$= R_G / \left[1 - \frac{R_2}{R_1 + R_2} \cdot A_v \right]$$

If $R_1 \ll R_2$ (as is usually the case), and $A_v \approx 1$, then $R_i \approx R_G$, but it must be kept in mind that the input impedance Z_i is determined by the input capacitance ($\approx C_{gd}$ as the next example will show) in parallel with R_G.

Example 7.7   The parameters of the FET shown in the simple source follower circuit of Fig. 7.16 are g_m = 2.0 mS, C_{gd} = 4 pF, C_{gs} = 3 pF. If R_G = 5 MΩ and the load resistor is 4.7 kΩ, determine the low frequency voltage gain and the upper 3 dB point of the amplifier.

At low frequencies the capacitances will have negligible effect, therefore using equation (7.8) and working in mS and kΩ, we have

$$A_v = \frac{g_m R_L}{1 + g_m R_L} = \frac{4 \times 4.7}{1 + (4 \times 4.7)}$$

$$= 0.95$$

At high frequencies the effects of C_{gd} and C_{gs} cannot be ignored. From the equivalent circuit of Fig. 7.19 we have

$$i_i = v_i j\omega C_{gd} + (v_i - v_o) j\omega C_{gs}$$

and so

$$Y_i = \frac{i_i}{v_i} = j\omega \{C_{gd} + (1 - A_v) C_{gs}\}$$

The input capacitance is therefore

$$C_i = C_{gd} + (1 - A_v) C_{gs}$$

Hence

$$C_i = 4 + (1 - 0.95)3 = 4.15 \text{pF}$$

Notice that this is little different than C_{gd} and is negligible relative to the input capacitance (the Miller capacitance) of a common-source amplifier where $A_v \gg 1$.

The upper 3 dB frequency is given by

$$\omega_2 = \frac{1}{C_i R_G} = \frac{10^{12}}{4.15 \times 5 \times 10^6}$$

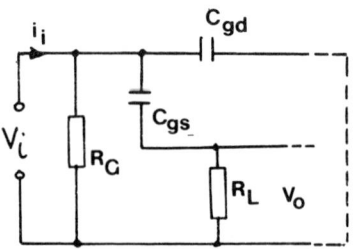

Figure 7.19

$$= 48193 \text{ rad/s or } f_2 = 7.67 \text{ kHz}$$

7.6.2   The emitter follower

The common-collector or emitter follower is the bipolar equivalent
to the FET source follower and is similar to the FET in that it provides improved output impedance for matching into low impedance
loads in return for a gain which is just below unity. The phase
conditions and the appearance of negative feedback are the same as
those for the source follower. The basic emitter follower circuit
and its equivalent are shown in Fig. 7.20(a) and (b) respectively.
Notice we use the common-emitter parameters.

From the equivalent circuit a set of defining equations may be
established; taking h_{re} to be negligible and ignoring the shunting
effect of the bias resistors we have

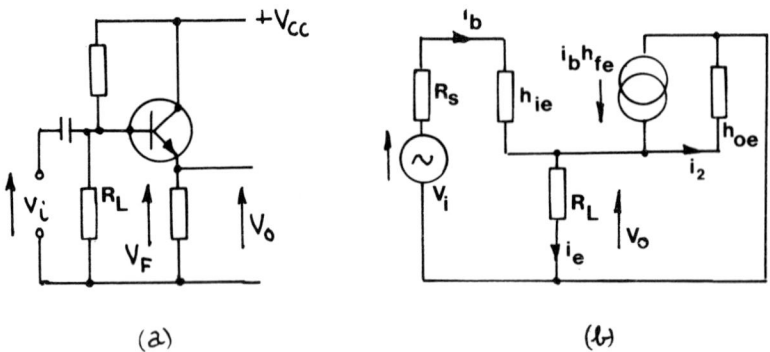

(a)   (b)

Figure 7.20

$$i_2 = h_{fe}i_b - i_e \qquad (i)$$

$$v_i = i_b(R_s + h_{ie}) + i_e R_L \qquad (ii)$$

$$0 = i_e R_L - \frac{i_2}{h_{oe}}$$

$$= i_e R_L - (h_{fe}i_b - i_e)/h_{oe}$$

$$= i_e(R_L + \frac{1}{h_{oe}}) - (h_{fe}i_b)/h_{oe} \qquad (iii)$$

Using these we can evaluate voltage and current gains, A_v and A_i respectively, and input and output impedances, Z_i and Z_o respectively.

An example will perhaps best illustrate this.

Example 7.8   A transistor with h_{ie} = 1.5 kΩ, h_{fe} = 100, h_{oe} = 33 μS is used as an emitter follower with an emitter load resistor R_L = 22 kΩ. The circuit is fed from a voltage source of internal resistance 50 Ω. Calculate A_i, A_v, Z_o and Z_i.

From the defining equations we have

$$i_2 = 100i_b - i_e \qquad (1)$$

$$v_i = 1.55i_b + 22i_e \qquad (2)$$

$$0 = (22 + 30)i_e - (30 \times 100)i_b$$

$$= 52i_e - 3000i_b \qquad (3)$$

Make a point that we are working in kΩ and that $1/h_{oe}$ = 30 kΩ. We can obtain A_i immediately from equation (3):

$$A_i = \frac{i_e}{i_b} = \frac{3000}{52} = 57.7$$

Using this result, we may write $i_b = i_e/57.7$ and then, since $A_v = v_o/v_i$

$$A_v = \frac{i_e R_L}{1.55i_b + 22i_e}$$

$$= \frac{R_L}{\frac{1.55}{57.7} + 22} = \frac{22}{22.03}$$

$$= 0.998$$

This result must, of course, be less than unity at all times. The value obtained for A_i, however, is roughly the same as that which would be obtained in common-emitter configuration. The FET has no meaningful current gain.

To obtain the output impedance, we replace v_o with an equivalent generator and short out the input source, see Fig. 7.21. Taking equations for the input and output loops yields

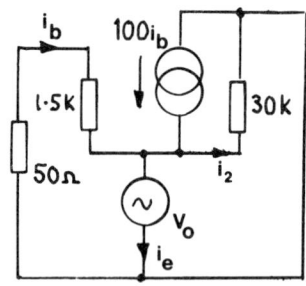

Figure 7.21

$$v_o = 1.5i_b + 0.05i_b = 1.55i_b$$

$$v_o = 3000i_b - 30i_e$$

The second of these derives from (3) earlier by setting $R_L = 0$ i.e. $v_o = i_2 \frac{1}{h_{oe}} = \frac{1}{h_{oe}}(h_{fe}i_b - i_e)$.

Then

$$v_o = \frac{3000}{1.55}v_o - 30i_e$$

Therefore

$$30i_e = 1935v_o - v_o$$

Therefore

$$Z_o = \frac{v_o}{i_e} = \frac{30}{1934} = 0.0155 \text{ k}\Omega$$

$$= 15.5\,\Omega$$

For the input impedance, we use equation (2):

$$v_i = 1.55i_b + 22i_e$$

But

$$i_e = 57.7i_b$$

Therefore

$$v_i = 1.55i_b + 22(57.7i_b)$$

$$= 1.55i_b + 1270i_b$$

Therefore

$$Z_i = \frac{v_i}{i_b} = 1271.5 \text{ k}\Omega \quad \text{or} \quad 1.27 \text{ M}\Omega$$

As we might have expected, the negative feedback has reduced the output impedance and increased the input impedance.

Both the source and the emitter follower circuits are essentially impedance transformers having the advantage over an ordinary wound transformer in that the impedance match between high and low impedances is accomplished without the reduction in voltage that a wound transformer would introduce. They are particularly useful in matching a high impedance source into a capacitive load, such as a long screened cable connection from the follower to a low impedance load. The shunting effect of the cable capacitance will have negligible effect, even at quite high frequencies, on the power delivered to the load.

7.6.3 The phase splitter

The so-called phase splitter is an amplifier, FET or bipolar, in which load resistors are placed in both the source (or emitter) and the drain (or collector) circuits. Outputs are then available at both terminals.

239

We consider here the FET phase splitter. Let the FET have drain and source load resistors, R_D and R_L respectively as shown in Fig. 7.22. Then if R_L is effectively decoupled to signal frequencies so that there is no feedback, the gain of the stage, as derived earlier by equation (5.15) is that of the common-source configuration. This is repeated here for convenience:

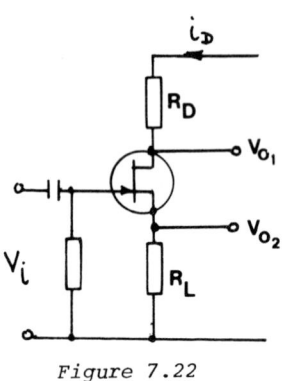

Figure 7.22

$$|A_{vi}''| = -g_m \frac{r_d R_D}{r_d + R_D} = \frac{-\mu R_D}{r_d + R_D}$$

since $\mu = r_d g_m$.

When the source resistor R_L is decoupled, feedback takes place and the effective input between source and gate is no longer v_i but $(v_i - v_F)$ where $v_F = i_D R_L$.
But

$$i_D = \frac{-\mu(v_i - i_D R_L)}{r_d + R_L + R_D}$$

and so

$$i_D(r_d + R_L + R_D + \mu R_L) = -\mu v_i$$

Therefore

$$i_D = \frac{-\mu v_i}{r_d + R_D + (1 + \mu) R_L}$$

For the output at the drain, $v_{o1} = i_D R_D$, hence the gain at this terminal with the feedback is

$$A'_{vd} = \frac{v_{o1}}{v_i} = \frac{-\mu R_D}{r_d + R_D + (1 + \mu) R_L} \tag{7.10}$$

For the output at the source, $v_{o2} = i_D R_L$, hence the gain at this terminal with the feedback is

$$A'_{vs} = \frac{v_{o2}}{v_i} = \frac{+\mu R_L}{r_d + R_D + (1 + \mu) R_L} \tag{7.11}$$

When $R_D = 0$, this last equation reduces to

$$\frac{+\mu R_L}{r_d + (1 + \mu) R_L}$$

which is the source follower case derived in equation (7.8).
When $R_L = R_D$

$$|A'_v| = \frac{\mu R_L}{r_d + R_L(2 + \mu)} = \frac{\mu R_D}{r_d + R_D(2 + \mu)}$$

and the outputs are equal. They are also less than unity and 180° out of phase. This circuit is a split phase inverter or 'phase splitter' and is commonly employed to act as a driver to a push-pull amplifier stage or in any system where equal and antiphase signals are required.

A couple of interesting points arise from the previous discussion. In deriving the gain of the circuit (at the collector) we did not use the general feedback formula. The gain at the source is obviously always less than unity, but the gain at the collector will be greater than unity with the proper selection of R_D. The feedback fraction

$$\beta = \frac{v_F}{v_{o1}} = \frac{i_D R_L}{i_D R_D} = \frac{R_L}{R_D}$$

and as an exercise you should be able to derive equation (7.9) by using the feedback formula.

The other point concerns the output impedance as seen at the drain. With R_L bypassed so that there is no feedback, this is r_d. With feedback applied, it is seen from equation (7.10) that the output impedance becomes $r_d + (1 + \mu)R_L$.

7.6.4   The bootstrap follower

This is a form of emitter follower but with a greatly increased input impedance. In the conventional emitter follower, although the input impedance of the device itself is very high, the shunting effect of the bias resistors on the input is often considerable, in typical cases presenting a shunt resistance of as little as a few thousand ohms.   This difficulty is not so pronounced in the source follower where the gate resistor is often several megohms in value and its shunting effect is not significant.

In the so-called bootstrap circuit, the normal bias resistors R_1 and R_2 are connected through R_3 to the base terminal (see Fig. 7.23); a capacitor C also connects their junction to the emitter. The proper bias conditions are maintained at the base by the combination of R_1, R_2 and R_3 and C couples v_o back to the junction of R_1 and R_2.   The emitter follower has $A_V \simeq 1$ and provides no phase shift between input and output.   Capacitor C is chosen so that its reactance is small at the signal frequencies and so there is virtually the same signal amplitude at both ends of R_3; hence R_3 draws only a very small current from the source and the bias chain is effectively isolated from signal input.

The factor by which the effective value of R_3 is increased above its actual value is clearly a function of A_V, since the signal voltage across R_3 is $(v_i - A_V v_i)$, and the current is therefore reduced by a factor $v_i/(v_i - A_V v_i)$.   For $A_V = 0.99$, the effective value of R_3 is increased 100 times; typically this may be of the order of one or two megohms.   The source follower with the split source load

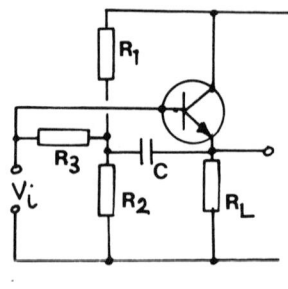

Figure 7.23

described under Fig. 7.18 is a form of bootstrap circuit.

7.7   MULTISTAGE FEEDBACK

Feedback may be applied to a single stage or over each of the individual stages making up a multistage amplifier; or it may be applied to
the multistage amplifier as feedback embracing two or more stages
in a single feedback loop. Feedback over single stages is known as
'local' feedback and its use is common, generally because of the simplicity of calculation and the easy avoidance of instability. Feedback embracing two or more stages in a single feedback loop is known
as 'overall' feedback. Is any advantage to be gained by using one of
these arrangements in preference to the other?

Suppose an amplifier has m identical stages each providing a gain
A. We shall not use a subscript here to help simplify the equations.
The total amplification A_t will be

$$A \times A \times A \times \ldots \ldots \text{ to m terms}$$

$$= A^m$$

Then

$$\frac{dA_t}{dA} = mA^{m-1} \tag{7.12}$$

This differentiation shows that the overall gain can change considerably for a small variation in the stage gain. For example, for
three stages each with a gain of 10, the overall gain is 1000. But
if a 10% change in the stage occurs, the total gain change is

$$dA_t = mA^{m-1}dA$$

$$= 3 \times 10^2 \times 0.1 = 30$$

which is 300 times the stage variation.

Suppose now that a multistage amplifier with n identical stages
each providing a gain A has feedback applied. Then

$$A' = \frac{A^n}{1 - \beta A^n} \tag{7.13}$$

and in order that the overall gain here will be the same as that of the amplifier having no feedback, $A_t = A'$ and so

$$A^m = \frac{A^n}{1 - \beta A^n}$$

From this we find

$$\beta = \frac{A^m - A^n}{A^{m+n}} \qquad\qquad (7.14)$$

Differentiating (7.12) and substituting the value for β from (7.14) yields

$$\frac{dA'}{dA} = nA^{2m-n-1}$$

and since $dA_t = mA^{m-1}dA$ from (7.12) we may write this last equation in the form

$$\frac{dA'}{dA} = \frac{nA^{2m-n-1}}{mA^{m-1}} \frac{dA_t}{dA}$$

which, by suitable rearrangement may be written as

$$\frac{dA'}{A'} = \frac{n}{m} \cdot \frac{1}{A^{n-m}} \cdot \frac{dA_t}{A_t} \qquad\qquad (7.15)$$

This equation enables us to compare the relative gain stability of feedback and non-feedback systems. An example will perhaps best illustrate this point.

Example 7.9   An amplifier without feedback has a stage gain of 40. A feedback amplifier having three stages is used to replace the single stage amplifier. What must be the feedback fraction of this amplifier? Compare the relative changes in the feedback amplifier with those in the no feedback amplifier resulting from variations in the stage gain A.

For the single stage amplifier $m = 1$, $A = 40$. The feedback amplifier has three stages ($n = 3$) and requires a total gain of 40. Hence, using (7.14) we have

$$\beta = \frac{40^1 - 40^3}{40^4} = -0.025$$

This same result could have been obtained directly from the relationship $\beta \approx 1/A$.

The ratio of the changes in the two amplifiers for a given variation in stage gain is

$$\frac{dA'/A'}{dA_t/A_t} = \frac{n}{m} \frac{1}{A^{n-m}}$$

$$= 3 \frac{1}{40^2} = \frac{1}{533}$$

The feedback amplifier is clearly much superior, as we should expect.

7.7.1   Local or overall?

In a multistage amplifier, feedback may be applied locally to each of the individual stages, or it may be applied overall. In practice a choice may have to be made between these two possibilities and there may well be a mixture of the two forms in some designs. We can, however, investigate which of the two might be better from the point of view of gain stabilization in the general sense.

Fig. 7.24(a) shows a multistage amplifier with local feedback applied over each stage. The gain of each stage is A, the feedback on each stage is β_1 and the overall gain is A_1'. Fig. 7.24(b) shows a multistage amplifier with overall feedback. There are n stages giving an overall gain of A^n without feedback, and a gain of A_2' with feedback, the feedback factor being β_2.

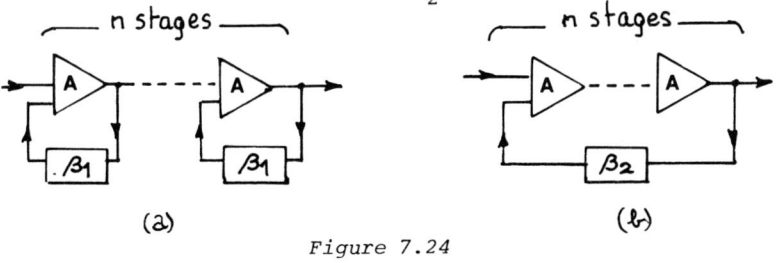

(a)   (b)

Figure 7.24

Then

$$A_1' = \left[\frac{A}{1 - \beta_1 A}\right]^n$$

$$A_2' = \frac{A^n}{1 - \beta_2 A^n}$$

Differentiating each of these we have

$$\frac{dA_1'}{A_1'} = \frac{n}{1 - \beta_1 A} \frac{dA}{A}$$

$$\frac{dA_2'}{A_2'} = \frac{n}{1 - \beta_2 A^n} \frac{dA}{A}$$

For both amplifiers to have the same gain, $A_1' = A_2'$. Therefore

$$(1 - \beta_1 A)^n = 1 - \beta_2 A^n$$

It then follows that

$$\frac{dA_2'/A_2'}{dA_1'/A_1'} = \frac{1}{(1 - \beta_1 A)^{n-1}} \qquad (7.16)$$

For $n = 1$, the denominator of (7.16) is unity, as expected. For $n > 1$ and for $(1 - \beta_1 A)$ large and positive i.e. considerable feed-back, then the fractional gain change dA_2'/A_2' is less than dA_1'/A_1'. Hence, overall feedback would seem to be best as far as gain stabil-isation is concerned. In practice, problems of stability limit the number of stages over which feedback can be usefully applied.

Further applications of negative feedback over a number of stages will turn up in the following chapter where operational amplifiers are discussed.

7.8   GAIN AND PHASE MARGIN

Example 7.5 illustrated how a change in the magnitude and phase of the feedback factor could turn a stable amplifier into an unstable one. In a properly designed feedback amplifier, it is necessary to provide a 'margin' of safety, as it were, between the operating values of gain and phase in the feedback loop and those values which would lead to the amplifier becoming unstable.

Gain margin is defined as the amount by which the mid-frequency loop gain of a stable feedback amplifier must be raised to just cause instability. An illustration is provided in Fig. 7.25.

Phase margin is defined as the extra phase change needed at unity loop gain, to just cause instability. This is illustrated in Fig. 7.26.

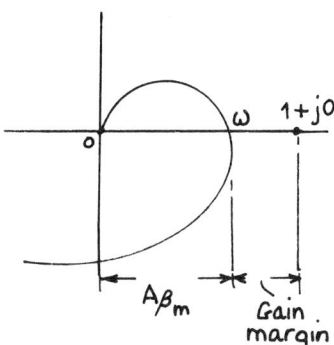

Figure 7.25

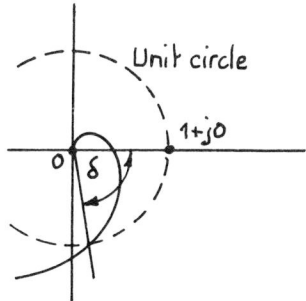

Figure 7.26

In Fig. 7.25, the loop gain at frequency ω is, say, $A\beta_m$ and this must be increased by a factor $1/A\beta_m$ before the gain becomes unity and the critical point is intersected. For example, if $A\beta_m = 0.1$, the loop gain may be increased by a factor of 10 before the circuit becomes unstable. A satisfactory gain margin will have a value between 0.1 and 0.3, corresponding to decibel gains of 10 and 20 dB respectively.

In Fig. 7.26, the angle δ represents the amount by which the phase angle can be changed without a change in the gain before instability occurs. This is the phase margin which is normally designed to lie between 30° and 60°.

7.9   A LABORATORY EXPERIMENT
This experiment investigates the behaviour of an emitter-follower not

only as a feedback amplifier but also as an impedance transformer.

Connect up the circuit of Fig. 7.27, using a transistor such as a BC107. The voltages indicated are measured with a high impedance a.c. voltmeter.

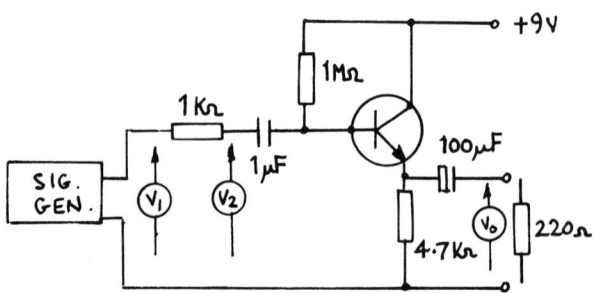

Figure 7.27

Set the frequency of the signal generator to, say, 500 Hz and adjust its output amplitude so that input voltage v_2 is at some definite value such as 0.1 V. Without the load resistor R_L connected, record the output voltage v_0; then connect R_L and readjust the signal generator output to restore voltage v_2 to its original value. Record voltages v_1^* and v_0^* (the asterisk indicates that the load is connected).

Retune the signal generator to 25 Hz and repeat the above procedure for a number of spot frequencies within the range 25 Hz to about 50 kHz. The input resistance $R_i = \dfrac{1000v_2}{v_1 - v_2}$ without the load and $\dfrac{1000v_2}{v_1^* - v_2}$ with the load connected. Similarly, the output resistance

$$R_o = \frac{220(v_0 - v_0^*)}{v_0^*} \, .$$

From your recordings plot a graph of input resistance against log frequency (without the load), and on the same axes the input and output resistances when the amplifier is loaded. What is the voltage gain of the amplifier? Can you deduce the current gain?

PROBLEMS 7

1. In a certain amplifier $A_v = 50\underline{/0^o}$. Find the magnitude of the gain when feedback is applied with $\beta = 0.05$, $\beta = 0.02$, $\beta = -0.02$.

2. Show that the gain A' of any amplifier with feedback relative to its gain A without feedback may be expressed as

$$A' = \frac{A}{1 - \beta A}$$

where β is the fraction of the output voltage fed back to the input.

Criticise the statement that this result applies to positive feedback and that if the feedback is negative, β is negative and therefore

$$A' = \frac{A}{1 + \beta A}$$

3. An amplifier has a voltage gain of -60 which varies over a range of ±5% of this value due to supply fluctuations. What will be the gain if n.f.b. with β = 0.06 is applied and what will be the percent variation in the gain?

4. An amplifier, A, in the absence of feedback has a gain which is liable to fall by 40% of its rated value as a result of supply variations. By the application of n.f.b. an amplifier is to be provided which has a rated gain of 100 and with the requirement that this gain should not fall below 99 using the amplifier A. Find the necessary gain of A and the feedback factor β.

5. An amplifier has an open-loop gain of 300 which is found to fall by 20% because of changes in the supply voltage. If the gain is to be stabilised so that it falls by only 1%, calculate the amount of n.f.b. required. If the original 3 dB bandwidth of this amplifier was 500 Hz to 50 kHz, calculate its bandwidth after the feedback is applied.

6. 'One of the chief reasons for the use of negative feedback in amplifier design is the elimination of instability, and to obtain the full benefits from such feedback a large amount must be used'.

Examine the contentions raised in this statement, commenting particularly on whether or not it is correct for multistage amplifiers.

7. The voltage gain of an amplifier without feedback is -100 -j80. Determine the magnitude and phase of the effective voltage gain of

the amplifier if the feedback circuit adds to the input signal a voltage that is 0.05 of the output voltage and lags behind it by 90°. Is this a case of positive or negative feedback?

8. The gain-frequency figures for an amplifier taken at four points in the frequency spectrum are

Freq.	10 Hz	1 kHz	20 kHz	100 kHz
Gain	100/180°	10^4/60°	600/-60°	10/-180°

If negative feedback is now applied to the amplifier, and assuming that the feedback path introduces no phase shift and that the feedback factor is independent of frequency and equal to -0.008, examine the stability of the amplifier at 10 Hz and 100 kHz. Calculate the new values of gain and phase angle at 1 kHz and at 20 kHz and comment on your solutions.

9. A student, when assembling what should have been the common-source voltage amplifier of Fig. 7.28, inadvertently omitted the source bypass capacitor C_s. By drawing a small-signal model of the circuit, derive an expression for the voltage gain v_o/v_i and deduce what effect the omission of C_s had on the gain if $R_s = 5/g_m$.

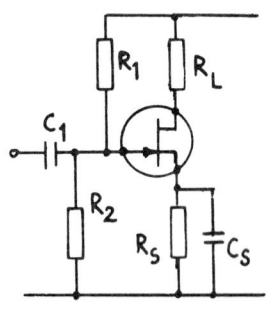

Figure 7.28

10. An amplifier has a voltage gain A_v and a fraction of its output is fed back in opposition to the input. If $\beta = 0.1/0°$, calculate the change (in dB) in the gain of the system if, because of supply variations or other causes, the value of A_v falls by 6 dB from an initial value of (a) 100/0°, (b) 100/90°.

11. A single stage RC-coupled amplifier in common-emitter configuration, has a mid-frequency voltage gain A_o = -50 and the 3 dB frequencies are f_1 = 25 Hz, f_2 = 15 kHz. A voltage-voltage feedback network is added to this amplifier in which the feedback fraction β = 0.05.

Assuming that β remains constant and is independent of frequency and that the phase shift is zero, sketch from a suitable table of related values a Nyquist plot of the circuit performance over an infinite frequency range and deduce that the amplifier is stable.

12. An amplifier has a gain of 57 dB. When n.f.b. is applied the gain falls to 27 dB. Show that the feedback factor β = 0.043.

13. Is oscillation possible in a two-stage RC-coupled common-emitter amplifier? Explain your answer.

14. Sketch the circuit diagram of an emitter-follower and describe how the circuit acts as an impedance transformer between a source and load.

What is the approximate power gain of an emitter follower having an a.c. input resistance of 100 kΩ and a load of 1 kΩ.

15. The a.c. circuit of an emitter follower is shown in Fig. 7.29. Draw an h-parameter equivalent circuit using EITHER common-emitter or common-collector parameters. Show by derivation of the expressions for input and output resistance, that the circuit may be used as an impedance transformer. What are the approximate values of the input and output resistance if R_L = 4.7 kΩ, R_S = 10 kΩ, h_{fe} = 50?

16. Using the equivalent circuit of the previous question, or other-

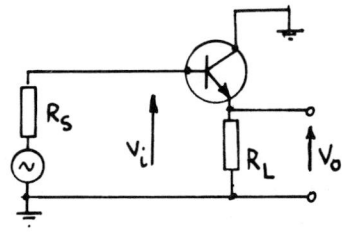

Figure 7.29

wise, show that as the load resistance R_L varies from zero to infinity, the input resistance of the stage varies from h_{ic} to $h_{ic} - h_{rc}h_{fc}/h_{oc}$.

17. If the response of an amplifier at low frequencies is expressed as

$$A_L = \frac{A_o}{1 - j[\frac{f_1}{f}]}$$

where A_o is the mid-frequency gain and f_1 is the lower 3 dB point, derive an expression for the lower 3 dB point of a feedback amplifier in terms of A_o, β and f_1'.

18. An amplifier having $A_o = 10^4$ has equal input and output resistances of 3 kΩ. If this amplifier is to be connected into a line having an impedance of 150 Ω resistive, what value of feedback factor must be applied to the amplifier to make proper matching possible? What will be the resulting gain of the amplifier? Suggest a suitable feedback arrangement.

19. The high frequency response of an amplifier is expressed by

$$A_H = \frac{A_o}{1 + j[\frac{f}{f_2}]}$$

where $A_o = 100$ and $f_2 = 1590$ kHz. Plot a graph of A_H/A_o against f on a logarithmic frequency scale (only two or three points are necessary). Now assume that a feedback loop with $\beta = -0.04$ is introduced; sketch the graph of A_H/A_o under this condition. Label both your graphs appropriately.

20. A 3-stage RC-coupled amplifier has identical stages each with $A_o = -10$. For each stage $R_i = 100$ kΩ, $R_o = 100$ Ω, $C_i = 0.1$ μF, and the coupling capacitors are each 10 μF. If overall n.f.b. is applied with $\beta = 0.01$, plot a Nyquist diagram and show that the amplifier appears conditionally stable but that a check on βA when $\underline{/\phi} = 0^o$ is necessary. By how much must the gain be increased to produce oscillation and at what frequency will this occur?

8 Direct-coupled and operational amplifiers

8.1   INTRODUCTION

Whilst all amplifiers inevitably exhibit a declining gain, and
associated phase shift, at high frequencies, this need not necessarily be so at low frequencies, and an amplifier with a flat response extending down to zero frequency is nowadays relatively
commonplace. Most integrated circuits contain such amplifiers where
they are usually known as 'operational amplifiers'. It is to be
appreciated that the circuits discussed so far have used capacitor
coupling between stages: these capacitors can transmit a.c. or time-
varying signals only and (besides attenuating the low frequencies)
act as a total block to direct current. Direct-coupled amplifiers,
or direct-current amplifiers as they are more usually called
(though this is a less technically descriptive name) have, as their
main application in life, the amplification of signals which vary
only slowly with time or go down to the d.c. level. Such amplifiers
will, of course, deal with a.c. signals perfectly normally, subject
only to the usual decline in gain at high frequencies.

8.2   THE DIRECT-COUPLED AMPLIFIER

It is possible to remove the coupling capacitors from an a.c.
coupled amplifier and replace them with direct connections. The
direct potential at each collector is then applied to the following
base. This is not an impracticable situation provided that the
emitter and collector potentials of the second stage are adjusted
to maintain the required operating voltages of that stage. However,
all such voltage levels depend ultimately upon the stability of the

supply voltage and any variations which may occur in the circuit component values or device parameters. Any such variations will be amplified as slowly changing d.c. signals, and when appearing at the output will be indistinguishable from the signal itself. Such a problem which is peculiar to direct-coupled amplifiers comes under the name of 'drift', the condition in which an output signal is obtained even in the absence of an input signal, or a gradual change occurs in the output level when the input signal is held constant. In bipolar transistor amplifiers particularly, drift depends upon

1. the collector supply voltage
2. the temperature dependent parameters of current gain, base-emitter voltage and leakage current.

The first of these possible causes can be eliminated by using a highly stabilised power supply, but the other factors are not so easily dismissed. An output other than zero will be obtained if a change in gain occurs, even if the input is zero. When the operating point of a stage varies as a result, for example, of a temperature variation, the output will change because in the absence of the coupling capacitors, any such operating point change will alter the operating points of all the following stages.

Leakage current is a problem because of its extreme sensitivity to temperature. Such currents set up in the system are amplified along with the required signal current. This problem is eased by the use of silicon devices and compensation is possible in the circuitry by the use of temperature sensitive resistors (thermistors) and diodes. We have already mentioned this in the chapter on power amplifiers. Both thermistors and diodes exhibit negative temperature coefficients of resistance and, when used in conjunction with 'ordinary' resistors have zero or positive coefficients, systems may be designed having negligible temperature sensitivity.

More serious is the variation in base-emitter voltage with temperature. As already noted, V_{be} changes by about 2 mV for each $^{\circ}C$ change in temperature, decreasing as the temperature increases. This may seem negligible but for a transistor with a gain, say, of 20, the output will drift by 40 mV for each degree increase in temperature just because the 2 mV has been treated by the transistor as an input signal. A term often used in connection with this respect is

254

the 'residual drift voltage', which is that voltage which when
applied to the input of the amplifier produces the SAME change in
the output of the amplifier as is produced by the temperature dependent variations in the transistor parameters.

8.2.1   Design fundamentals

Let us discuss the direct-coupled amplifier of Fig. 8.1. This diagram is not a simplification but is quite feasible as a working circuit. It would, however, have a number of practical limitations.

Because the coupling capacitor is absent, the direct potential
at the base of transistor T_2 is at the same level as the collector
of T_1. Clearly, to maintain the required operating conditions on T_2
something must be done to bring its emitter potential up to a value
slightly below its base level so that the correct base-emitter forward bias is obtained. When this has been done, the collector voltage of T_2 will be greater than the collector voltage of T_1 (why?), so
for a number of such stages in cascade, the collector voltages will
become even larger as successive stages are reached. The d.c. voltage level at the output terminals will, as a consequence, be greater
than the d.c. level at the input. So when the amplifier is inserted
between a source and an output load, the d.c. level is shifted. In
many cases this may be unimportant since only voltage changes are
of interest, but there is a problem in that the movement of the
operating point along the load line eventually restricts the possible
signal excursions of the later stages. The difficulty can be mini-
mised by using the circuit system shown in Fig. 8.2. Here the first
and third stages use n-p-n transistors, while the second stage is

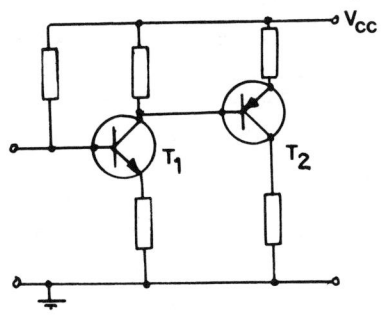

Figure 8.1

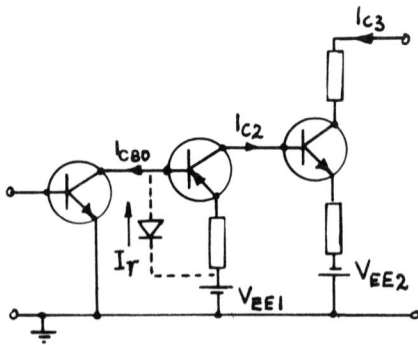

Figure 8.2

p-n-p. The theory behind this arrangement can be deduced by consider-ing the base and collector current directions. Assuming no drive to the base of the first stage transistor T_1, the collector current will be solely I_{CEO} flowing in the direction indicated. This current will be I_B for transistor T_2 and since this is a p-n-p device, the cur-rent direction will be correct for Class-A working. Let the current gain of T_2 be h_{FE2} so that its collector current will be $I_{C2} = h_{FE2} \cdot I_{CEO1}$. Again, the direction of I_{C2} will be correct as the base current of T_3 and the collector current of this third stage will be

$$I_{C3} = h_{FE3} I_{C2} = h_{FE2} \cdot h_{FE3} \cdot I_{CEO1}$$

It is clear that the sensitivity of such a circuit to leakage current and hence to ambient temperature variations is very great. The output signal is dependent upon the leakage current of the first stage and even if silicon devices are used, a variation of some 2-3% per degree C in that current might well be expected. Temperature compensation is necessary for such an amplifier and a diode is often used for this purpose because the temperature coefficient of resis-tance of the diode is of the same order as the rate at which the transistor leakage I_{CBO} varies with temperature. The diode (shown dotted) in Fig. 8.2 serves such a purpose. The diode is reverse biased and the reverse saturation current is indicated by I_r. Since V_{EE2} is the only voltage source at this point, $I_{B2} = -I_r + I_{C1}$, and if I_{B2} is to be maintained constant although I_{C1} (which may be I_{CEO}) may vary with ambient temperature, we require that

$$\frac{dI_{B2}}{dT} = 0$$

So for ideal compensation

$$\frac{dI_r}{dT} = \frac{dI_{C1}}{dT}$$

The circuit of Fig. 8.2 uses separate supplies to provide the correct collector to emitter voltages for each stage and this is not a convenient arrangement. Figure 8.3 shows a practical arrangement using one n-p-n and one p-n-p transistor and a balanced supply. The emitter of T_1 is held at a potential slightly below earth (about -0.6 V) and is correctly biased if its base is returned to earth (0 V) by way of R_1. The output level of T_1 is clearly above earth and this is the potential at the base of T_2. To get the potential at the collector of T_2 at zero volts which is necessary if the output voltage is to be zero for zero input to T_1, T_2 is connected to work between the $+V_{cc}$ and $-V_{cc}$ supply rails. The voltage drop across R_5 must consequently be equal to $-V_{cc}$.

The purpose of the variable resistor R_6 is to make adjustable the potential at the emitter of T_1 and hence the collector potential of T_1. If zero input does not result in zero d.c. output, the amplifier is said to exhibit an 'off-set voltage' and R_6, the 'off-set null'

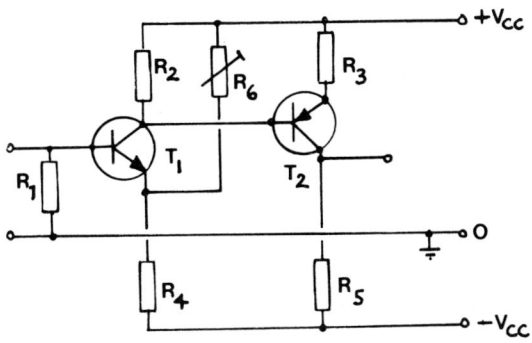

Figure 8.3

control is adjusted to provide the desired zero output.

Example 8.1   Given that R_2 = 3.3 kΩ, R_3 = R_4 = 1 kΩ, R_5 = 5.6 kΩ and R_6 = 1.5 kΩ and that $+V_{cc}$ = $-V_{cc}$ = 10 V in the circuit of Fig. 8.3, calculate the potential levels present in the circuit and estimate the voltage gain of the amplifier.

The collector of T_2 must be at zero d.c. voltage, hence the voltage drop across R_5 must be 10 V. So I_C for I_2 = 10/5600 A = 1.78 mA and this current will develop a voltage of 1.78 V across the 1 kΩ emitter resistor R_3. The emitter potential will therefore be (10 - 1.78) = 8.22 V above earth.

The base of T_2 must be held about 0.6 V negative with respect to its emitter to maintain these other potentials, hence the base potential of T_2 is (8.22 - 0.6) = 7.62 V above earth. This level is suitable as the quiescent collector voltage of T_1 and by adjustment of R_6 such a level can be readily achieved.

In the absence of further information, the voltage gain of each stage must be approximated as $A_v \simeq R_L/R_E$. For the first stage, R_L = 3.3 kΩ and R_E = $R_4 \| R_6$ = 0.6 kΩ. Hence

$$A_{v1} = \frac{3300}{600} = 5.5$$

$$A_{v2} = \frac{5600}{1000} = 5.6$$

Hence the overall gain is approximately 5.5 × 5.6 ≃ 31.

This example should have illustrated the usefulness of complementary pairs of transistors in d.c. amplifiers. It should also have made clear that the zero output signal and the input signal can be the same only if an even number of stage is used in the amplifier.

8.2.2   The Darlington connection
The Darlington or composite transistor circuit shown in Fig. 8.4 is an interesting and widely used configuration involving d.c. coupling between two (or three) transistors. The emitter of T_1 is directly connected to the base of T_2 and the collectors of both devices share a common load. As for a single transistor, only three terminals are presented to the external circuit.

In this circuit the emitter current I_{E1} of T_1 is the base current

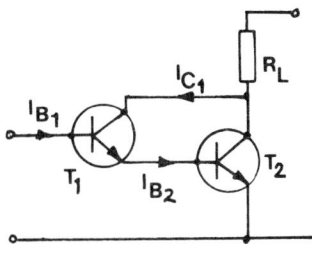

Figure 8.4

of T_2. If R_L is small, $I_{E1} = (h_{FE1} + 1)I_{B1}$ and $I_{C2} = h_{FE2}I_{B2}$. Then the overall gain is

$$\frac{I_{C2}}{I_{B1}} = \frac{h_{FE2}(h_{FE1} + 1)I_{B2}}{I_{E1}} = h_{FE2}(h_{FE1} + 1)$$

since $I_{E1} = I_{B2}$. Since also $h_{FE1} \gg 1$, the overall gain approximates to $h_{FE1}h_{FE2}$.

As for any other d.c. amplifier, leakage current is also amplified in the Darlington configuration and resistors are used to reduce the effect of such current; see Fig. 8.5. From the relationship

$$S = \frac{\delta I_C}{\delta I_{CBO}}$$

the leakage current from transistor T_1 is divided between resistor R_2 and the input of transistor T_2. This current is $S_1 I_{CBO1}$. Also part of the leakage current I_{CBO2} of T_2 may flow through R_2 to reduce the stability factor S_2 of T_2. As R_3 is unbypassed, this resistor should

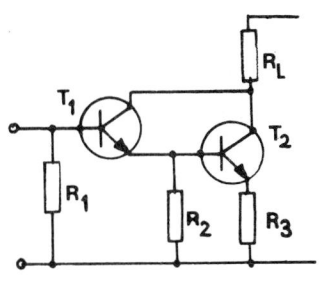

Figure 8.5

259

be small in comparison with the load R_L if a good voltage gain is required. The signal gain is in fact reduced through the shunting effect of the resistors. If the resistors R_1 and R_2 are replaced by diodes in reverse connection and the reverse saturation currents of these diodes are equal to the respective leakages, the current stability factor of each transistor is unity. Additionally, because of the high reverse resistances of diodes, the current gain approaches the theoretical $h_{FE1}h_{FE2}$ very closely.

Darlington pairs and triples are available in a single package and current gains from 10,000 to 20,000 are easily obtainable.

8.3   THE DIFFERENTIAL AMPLIFIER

The best way of getting over the drift problem in d.c. amplifiers is to use some sort of balanced amplifier where drift variations in one part of the circuit are nullified by equal and opposite changes in another part. This condition is obtained in the so-called 'differential' (or 'difference') amplifier which finds almost universal application in integrated operational amplifiers.

Consider the circuit shown in Fig. 8.6 which illustrates the basic form of a differential amplifier. The name stems from the fact that the input signal v_i is balanced with respect to the earth line so that the two base inputs are antiphase to each other; as the forward bias of transistor T_1 is increased, the forward bias of transistor T_2 is reduced, and conversely. Assuming that the transistors are perfectly matched (as is closely possible on an integrated circuit chip where the corresponding elements are fabricated simultaneously) the resulting current through the common emitter resistance R_E (the tail resistance) remains constant. Hence the voltage drop across R_E is constant and no gain reduction occurs through the presence of this resistor, that is, it appears to the signal to be perfectly bypassed.

Like the input signal, the output signal is balanced with respect to earth and so may feed into a balanced load or a further balanced amplifier. Since the collector voltages swing in opposite directions, the output voltage taken between collectors for antiphase inputs is large. This is known as the 'differential mode' of operation. The main advantage of this kind of amplifier is not so much its operation

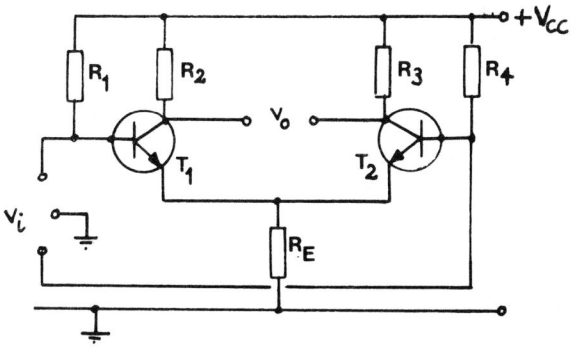

Figure 8.6

in differential mode but its behaviour in 'common mode' where in-phase signals are applied to the two input bases. The collector voltages now swing in the same direction and the output signal is, for perfect balance, the zero difference between the two collector potentials. Such in-phase signals include the thermally generated leakage currents, hence if both input sections drift to the same extent, there is no net drift in the output.

An analysis of the circuit may be made by assuming conditions in the common-mode and differential states; first that the input signals are identical, that is, $v_1 = v_2$ in Fig. 8.7(a) and second, that the input signals are different, that is $v_1 > v_2$ or $v_2 > v_1$.

With $v_1 = v_2$, the circuit behaves basically as an emitter follower and the voltage across R_E follows the input voltage. The total current in R_E is v_1/R_E and assuming perfect matching this current will be shared equally between T_1 and T_2. That is, taking $i_c \approx i_e$

$$\frac{v_1}{R_E} = i_{c1} + i_{c2} = 2i_{c1}$$

Now

$$v_o = -i_{c1}R_L = -v_1R_L/2R_E$$

and the common-mode gain

$$A_{vc} = -\frac{R_L}{2R_E} \tag{8.1}$$

261

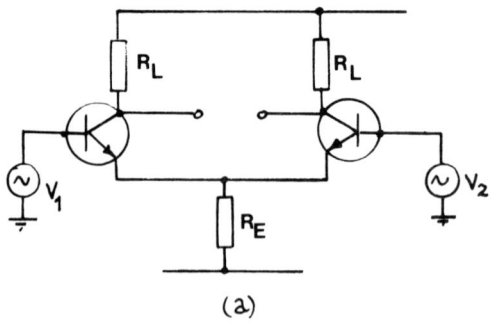

(a)

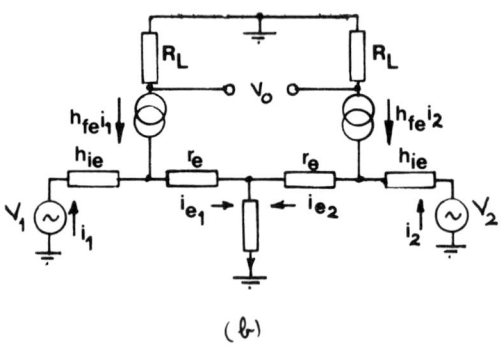

(b)

Figure 8.7

Clearly, the greater the value of R_E compared with the collector load R_L, the smaller the gain. In theory for zero gain, R_E must be infinite; in practice it must be large compared with R_L. If a passive component (resistor) is used, unrealistic values of V_{cc} may be called for to obtain the desired emitter current; a better solution (used exclusively in integrated amplifiers) is to replace R_E by a constant current source. We shall return to this aspect very shortly.

Now assume a differential input, that is, with a different signal on the base of T_1 from that on the base of T_2, see Fig. 8.7(a). We cannot assume that the emitters are connected together with negligible resistance as we did for the common-mode condition as this would imply that the inputs were identical. Each transistor exhibits a dynamic emitter resistance r_e. Assuming that $1/h_{oe} \gg R_L$ and that $R_E \gg r_e$, we have from the equivalent circuit of Fig. 8.7(b)

$$v_1 = i_1 h_{ie} + I_E R_E + i_1 r_e$$

$$v_2 = i_2 h_{ie} + I_E R_E + i_2 r_e$$

$$I_E = i_{e1} + i_{e2}$$

where

$$i_{e1} = (h_{fe} + 1)i_1 \quad \text{and} \quad i_{e2} = (h_{fe} + 1)i_2$$

Then

$$i_1 = \frac{1}{2} \cdot \frac{V_1 - V_2}{h_{ie} + (h_{fe} + 1)r_e}$$

$$i_2 = \frac{1}{2} \cdot \frac{V_2 - V_1}{h_{ie} + (h_{fe} + 1)r_e}$$

The voltage BETWEEN the collectors is therefore

$$v_o = h_{fe} R_L i_1 - h_{fe} R_L i_2$$

$$= h_{fe} R_L (i_1 - i_2)$$

Substituting for i_1 and i_2 obtained above we have for each collector relative to earth

$$v_o = \frac{1}{2} \frac{(V_1 - V_2)h_{fe} R_L}{h_{ie} + (h_{fe} + 1)r_e} \tag{8.2}$$

Thus the amplifier responds to the difference in voltage between the two inputs. Note that if $V_1 > V_2$ the output is negative, and if $V_2 > V_1$ the output is positive. Input 1 is referred to as the 'inverting' input and input 2 as the 'non-inverting' input. The differential voltage gain A_{vd} is given by

$$A_{vd} = \frac{v_o}{V_1 - V_2} = \frac{1}{2} \frac{h_{fe} R_L}{h_{ie} + (h_{fe} + 1)r_e}$$

and taking $(h_{fe} + 1) = h_{fe}$, $h_{fe} r_e \gg h_{ie}$

$$A_{vd} \simeq \frac{-R_L}{2r_e} \simeq -\frac{R_L g_m}{2} \qquad (8.3)$$

since $r_e = 1/g_m$.

So the effective transconductance of the differential amplifier is one-half the transconductance of either of its input transistors.

8.3.1 Common-mode rejection ratio

The factor of ½ which appears in equations (8.2) and (8.3) comes about because, when the output is considered from either collector relative to earth, the output is 'unbalanced' and the voltage gain between collectors is decreased by a factor of 2. The differential amplifier does not have to be balanced; both input and output can be in unbalanced form as Fig. 8.8 illustrates. Here the base of T_2 is earthed (input $v_2 = 0$) and the output signal is derived between the collector and earth of this transistor. Hence T_2 is working in common-base configuration, with T_1 in common-colector (emitter follower).

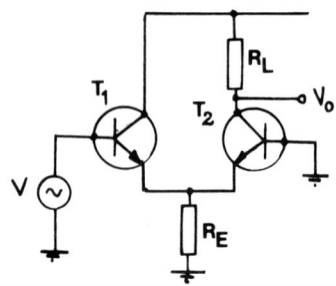

Figure 8.8

Although for certain reasons it is not entirely satisfactory to take a single-ended output from one side of a differential amplifier (without certain circuit modifications) we can nevertheless give a reasonable analysis of the circuit as it stands. The unbypassed R_E does not lead to a loss of gain as might at first be suspected because the common-base input impedance of T_2 is in parallel with R_E and h_{ib} is usually of the order of 20-30 Ω, very much smaller than R_E. The input resistance of T_1 is then $h_{ie} + (h_{fe} + 1)h_{ib} = 2h_{ie}$ if the transistors are identical.

When common-mode signals are applied to the unbalanced amplifier the output cannot be zero (as is theoretically possible in a balanced system) because the output voltage is referenced to V_{cc} or earth. However, unlike the differential mode, h_{ib} of T_2 is not in parallel with R_E when in-phase signals are applied and the amplifier gain may be small. Clearly we are not talking about externally applied signals since one base is earthed; the 'signals' concerned are those internally produced, being either I_{CBO} or δV_{BE} whether the input is balanced or unbalanced.

Ideally, the differential amplifier responds only to the difference voltage $(v_1 - v_2)$ and not to the common-mode voltage, but any practical system produces some common-mode output. This gives us a criterion for assessing the quality of a differential amplifer: the measurement of the common-mode rejection ratio (CMRR) which is defined as the ratio of the difference voltage gain to the common-mode voltage gain:

$$CMRR = \frac{\text{differential mode gain } A_{vd}}{\text{common mode gain } A_{vc}}$$

But

$$A_{vd} = \frac{R_L g_m}{2} \quad \text{and} \quad A_{vc} = \frac{R_L}{2R_E}$$

Therefore

$$CMRR = \frac{R_L g_m}{R_L/R_E} = g_m R_E = \frac{h_{fe}}{h_{ie}} R_E$$

Therefore, a high CMRR is obtained (for a given gain) when R_E is large in comparison with $h_{ie}/h_{fe} = h_{ib}$. This ties in with our earlier deductions.

Example 8.2   The differential gain of an amplifier is 20,000 and the common-mode gain is 5. What is the CMRR of this amplifier?

$$CMRR = \frac{20000}{5} = 4000$$

or, as it is usually expressed in dB

$$CMRR = 20 \log 4000 = 72 \text{ dB}.$$

This is a not unreasonable figure, although modern amplifiers have ratios of 80-100 dB.

Example 8.3   In the circuit of Fig. 8.8, R_L = 1 kΩ, h_{ie} = 1.5 kΩ, h_{fe} = 80 (for both transistors). Analyse the circuit and obtain an estimate of the overall voltage gain.

As mentioned above, the circuit is an emitter-follower directly coupled to a common-base amplifier. The voltage gain of an emitter-follower is approximately unity but this is not so in the present circuit. The signal input is effectively applied to two forward biased diodes in series, hence the voltage at the emitters is about one-half of v_1 and, for identical transistors, the effective voltage gain of T_1 is 0.5.

The input resistance of T_2 is clearly very low; it is in fact h_{ib} = $h_{ie}/(h_{fe} + 1)$ = 1500/81 = 18.5 Ω. The voltage gain of T_2 is about R_L/h_{ib} = 1000/18.5 = 54, hence the overall voltage gain is about 0.5 × 54 = 27.

8.3.2   Constant current source

As we have seen already, CMRR can be improved if R_E is increased but there is a practical limit of a few thousand ohms to the value of an ordinary resistor when used for R_E. The effective value of R_E can be enormously increased and yet allow the desired value of emitter current to flow (without using a high potential on V_{cc}), if a constant current source replaces R_E.

Such a source can be obtained from a transistor; if then our differential amplifier has R_E in the tail replaced by a transistor in common-emitter mode as Fig. 8.9 shows, the problem is overcome. The collector current of T_3 is practically constant and is maintained so by the stabilising bias circuit made up of R_1, R_2 and R_E'. The effective value of R_E is then given approximately by $1/h_{oe}$ of T_3 which will be of the order of 100 kΩ or more. So the CMRR is high, the amplifier has constant emitter currents, and the operating point is held constant. A diode can be usefully included in series with R_1 to provide temperature compensation.

The introduction of a constant current source in the emitter lead enables the operation of the amplifier to be summed up in a few words. Since i_{c1} + ic_2 is constant, any increase in i_{c1} results in

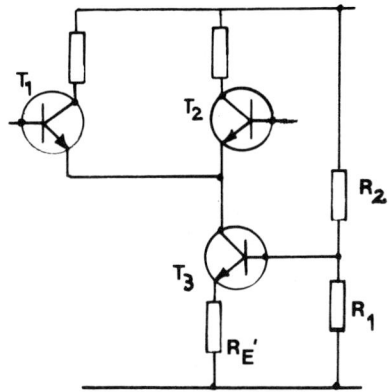

Figure 8.9

a decrease in i_{c2} and the collector voltage of T_2 (v_{c2}) increases. The output which is $v_{c2} - v_{c1}$ (v_{c1} is constant in the present circuit but varies in a balanced system) responds to the difference in the inputs, $v_2 - v_1$. If v_1 and v_2 both increase due to a common signal, T_1 and T_2 would have to respond in identical fashion but i_{c1} and i_{c2} cannot change because their sum must be constant. Therefore, if $v_1 = v_2$ no output is obtained.

Example 8.4   A transistor amplifier has a current-stabilising transistor in the emitter circuit as shown in Fig. 8.9. What is the CMRR if $h_{ie} = 1.2$ kΩ, $h_{fe} = 150$ for transistors T_1 and T_2 and $h_{oe} = 25$ μS for transistor T_3?

$$CMRR = \frac{h_{fe}}{h_{ie}} \cdot R_E$$

Here we may take

$$R_E = 1/h_{oe} = \frac{10^6}{25} = 40 \text{ k}\Omega$$

Then

$$CMRR = \frac{150 \times 40 \times 10^3}{1200}$$

$$= 5000$$

$$= 20 \text{ lg } 5000 = 74 \text{ dB}$$

8.4   THE OPERATIONAL AMPLIFIER

Not so many years ago the operational amplifier was a specialised
and very expensive piece of equipment whose original role was, and
still is in many applications, to perform the mathematical operations
of multiplication, division, differentiation and integration. These
early operational amplifiers used thermionic valves and were of
necessity very bulky articles. There was also the problem of heat
dissipation and ventilation. With the advent of the transistor, such
problems disappeared. The advantage of small bulk, low power con-
sumption, reliability and ruggedness, plus the reduction in cost,
took the operational amplifier out of the laboratory and into an
almost limitless variety of everyday signal processing, instrumentation and control applications. Although the early solid state
designs using discrete transistors suffered from temperature effects,
this could be overcome by a careful matching of pairs of transistors
in a differential configuration. In the modern integrated circuit
form, the corresponding elements are fabricated simultaneously in
very close proximity on the same base chip and are practically identical in all respects. The operational amplifier therefore developed
as a two-input terminal device and is basically a high performance,
directly-coupled amplifier capable of very stable operation over a
range of frequencies including zero.

8.4.1   The ideal model

The circuit analysis of the operational amplifier can be reduced for
most purposes to a few simple restraints. It is not necessary to
know the full internal circuit details of the amplifiers in order to
use them, but it is necessary to know the facilities provided by
them and the terms used to specify their performance.

As mentioned above, the operational amplifier is a high gain,
d.c. coupled unit with a frequency range extending down to zero. The
output is normally single-ended while the input can be either single-
ended or differential. For single-ended input, one of the differential inputs is earthed.

An idealised model of the amplifier is useful in analysing feedback circuits and the characteristics and functions performed by the
amplifier are normally determined by the form of external feedback

268

applied. The idealised characteristics of such a model are assumed to be:

Infinite voltage gain

Infinite input impedance

Infinite bandwidth

Zero output impedance

Zero output when the inputs are identical

When these ideal characteristics are incorporated (and they can be closely approximated in standard commercially available units) the circuit model of the amplifier reduces to that shown in Fig. 8.10. In developing an analysis, this circuit model will be used.

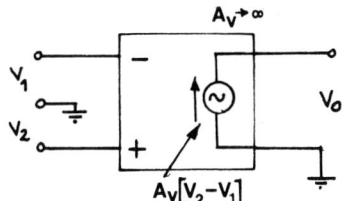

Figure 8.10

8.4.2   The inverting amplifier

The circuit arrangement of Fig. 8.11 belongs to the general category known as 'inverting' circuits. The common feature of these circuits is that the non-inverting input is connected to earth. Feedback and input networks are attached to the inverting input terminal. In analysing the circuit, we assume a gain A_V which is subsequently allowed to approach infinity. Since the amplifier has an extremely large input impedance, the input current is negligible and con-

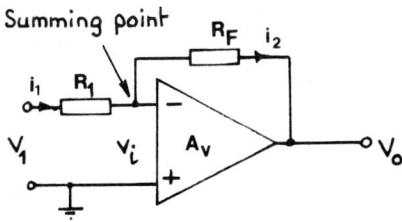

Figure 8.11

sequently resistors R_1 and R_F carry equal currents; hence $i_1 = i_2$
or

$$i_1 = \frac{v_1 - v_i}{R_1} = \frac{v_i - v_o}{R_F} = i_2 \qquad (i)$$

The amplifier gain

$$A_v = -\frac{v_o}{v_i} \text{ (inverting)} \qquad (ii)$$

Solving for v_i and substituting into (i) yields

$$\frac{v_i + \dfrac{v_i}{A_v}}{R_1} = \frac{-\dfrac{v_o}{A_v} - v_o}{R_F} \qquad (iii)$$

Now for $A_v \to \infty$ equation (iii) approaches the limit

$$\frac{v_o}{v_i} = -\frac{R_1}{R_F}$$

or

$$v_o = -\frac{R_1}{R_F} v_i \qquad (8.4)$$

Note that the gain includes the sign inversion and that its magnitude is determined solely by the ratio of the external resistors, that is, the gain of the circuit is independent of the voltage gain of the amplifier as long as this is very large. This is a typical result of the application of heavy negative feedback. Another important point is that for an ideal amplifier, any current arriving at the inverting terminal must flow through the feedback resistor R_F. If there were a number of signal sources connected to this terminal instead of only one by way of resistor R_1, the sum of the several currents arriving there would flow through R_F. For this reason, the inverting terminal is known as the 'summing point'. The summing point voltage, v_i, approaches zero as A_v approaches infinity, that is

$$v_i = -\frac{v_o}{A_v} \to 0 \quad \text{as} \quad A_v \to \infty$$

This condition is described by referring to the summing point as a 'virtual earth' since it is held at virtually zero or earth potential irrespective of the values chosen for R_1 and R_F. With this point at earth, the current through R_1 is $i_1 = v_1/R_1$ and therefore quite independent of R_F. Since negligible current flows into the amplifier, we may think of the input circuit as a source of current which must flow into R_F. Since one end of R_F is at earth potential, the other end must be at a potential of $-i_1R_F = V_o$. Therefore the input is isolated from the output. This leads to the simple model for an inverting amplifier shown in Fig. 8.11.

The feedback element represented by R_F need not be a resistor but can be any impedance Z. Indeed, it need not be a linear element but can be any element or group of elements, for which there is a linear or non-linear relationship between short-circuit current and terminal p.d. In the same way the input circuit may consist of general impedances. Regardless of the complexity involved in the input and feedback networks, the same principles will be found to hold; the summing point is a virtual earth and the current flowing into the summing point from the input must flow out into the feedback circuit.

8.4.3 The summing amplifier

If the basic inverting circuit is modified to accept additional signal sources and summing resistors are added, the amplifier performs addition. In Fig. 8.12, the summing point is at virtual earth and the currents through the input resistors are

$$i_1 = \frac{v_1}{R_1} \qquad i_2 = \frac{v_2}{R_2} \qquad i_3 = \frac{v_3}{R_3}$$

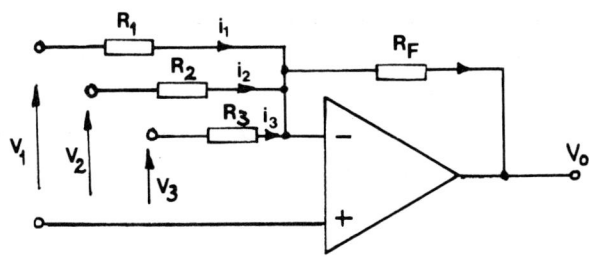

Figure 8.12

All these input signal currents flow through R_F, generating an output voltage

$$v_o = -iR_F = -(i_1 + i_2 + i_3)R_F$$

$$= -v_1 \cdot \frac{R_F}{R_1} - v_2 \cdot \frac{R_F}{R_2} - v_3 \cdot \frac{R_F}{R_3}$$

As the circuit model illustrates, the circuit functions as a scaling adder if all three input resistors are equal. If differing values are used, v_o is a weighted average of the input. Hence, for $R_1 = R_2 = R_3$

$$v_o = -\frac{R_F}{R_1}(v_1 + v_2 + v_3)$$

and for $R_1 = R_2 = R_3$

$$v_o = -R_F\left[\frac{v_1}{R_1} + \frac{v_2}{R_2} + \frac{v_3}{R_3}\right]$$

The circuit is therefore an 'analogue' adder, and the input resistances seen by the signal sources are R_1, R_2 and R_3 respectively.

8.4.4   The non-inverting amplifier

The inverting circuit just discussed can be realised with either single-ended or differential-input amplifiers. Non-inverting amplifiers in general require a differential input. The basic non-inverting circuit is shown in Fig. 8.13. Here the signal is applied to the non-inverting terminal (+) and the output (or some fraction of

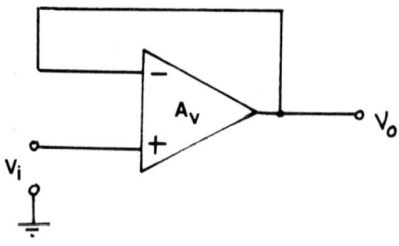

Figure 8.13

272

it) is fed back to the inverting (-) terminal. The equations describing the circuit are very elementary:

$$v_o = A_v(v_i - v_o)$$

so that

$$v_o = \frac{v_i}{1 + 1/A_v}$$

As $A_v \rightarrow \infty$, $v_o \rightarrow v_i$ and the circuit gain $\rightarrow 1$.

The output voltage will therefore always take on the value required to drive the signal between the inverting and non-inverting
inputs towards zero. This is a 'voltage-follower', a more sophisticated version of the emitter- or source-follower. Notice that the
gain is positive.

Since negligible current flows into the + input, the input impedance is very high; also, since no current flows through the feedback loop, any arbitrary (but finite) resistance may be placed in the
feedback loop without changing the properties of the idealised circuit. Like the emitter-follower, the voltage follower operational
amplifier is used as a buffer or impedance matching device. Used as
a 'power' amplifier, the voltage follower will permit a source of low
current capacity to drive a heavy load.

The voltage follower is, of course, a special and very useful
case of the non-inverting amplifier in general. In this general
form, shown in Fig. 8.14, the signal is applied to the + terminal and
a fraction of the output is fed back to the - terminal. From simple
feedback theory we have

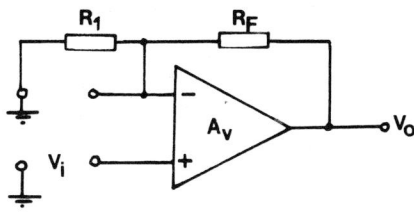

Figure 8.14

$$\beta = \frac{R_1}{R_1 + R_F}$$

and for A_V very large $A_V' \rightarrow \frac{1}{\beta} \rightarrow \frac{R_1 + R_F}{R_1}$

The overall gain is now greater than unity, is positive and is determined solely by the values of R_1 and R_F. Furthermore, as with the
emitter- and source-followers, the input impedance is large.

As for the inverting circuit, the feedback network need not be
a simple resistance divider but may be a combination of linear and
non-linear elements selected to provide a desired characteristic.
Example 8.5   It may be necessary for the amplifier of Fig. 8.14 to
handle only a.c. signals. Sketch a circuit diagram showing how this
could be done and comment on the arrangement.

A.C. coupling may be necessary in a particular case if the signal
source does not provide a d.c. path to earth or if a d.c. component
in the input signal must be isolated from the amplifier input. In
such cases the circuit of Fig. 8.14 must be modified to include a
coupling capacitor.

Figure 8.15 shows a suitable arrangement; a resistor R_3 is connected from the inverting input to earth and the coupling capacitor
is placed in series with the input from the source. This is ordinary
RC coupling.

A problem arises in that R_3 shunts the input and so nullifies the
inherently high input resistance of the amplifier. Clearly, to compensate for this, R_3 should be made as large as possible, but there
is a limit to this if a small offset voltage at the output is to be

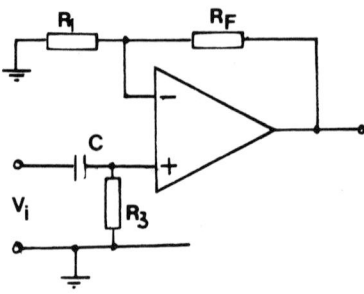

Figure 8.15

obtained. Since the coupling is a.c., a coupling capacitor may be connected in series with the output so that any offset is not passed on to the following stages. The value of R_3 may then be raised to a general value of 1 or 2 MΩ.

Alternately, a capacitor may be connected in series with R_1 in the feedback loop. R_3 can then be made as high as R_F (which may be several megohms) without a serious offset voltage developing. The point to be made about this arrangement is that the closed loop gain from the d.c. standpoint remains at unity no matter how high the a.c. gain is set.

8.4.5   The differential mode

The summing point principle applies, of course, to the differential mode of operation of an operational amplifier. The circuit is shown in Fig. 8.16. This circuit is widely used in instrumentation and, as we have seen, has the advantage of discriminating against d.c. variations, drift and noise input, and responds only to significant signal variations.

Since the amplifier draws negligible input current we have

$$v_n = i_2 R_3 = \frac{v_2}{R_2 + R_3} \cdot R_3$$

$$\frac{v_1 - v_p}{R_2} = \frac{v_p - v_o}{R_F}$$

Also, as $A_v \rightarrow \infty$, $v_p = v_n$

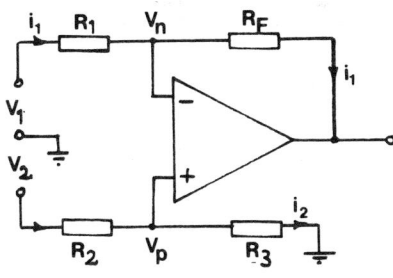

Figure 8.16

Substituting this relation into the previous two equations and solving for v_0 gives us

$$v_0 = - \frac{R_F}{R_1} (v_2 - v_1)$$

which recapitulates our analysis of the differential or difference amplifier whose gain is determined solely by the ratio of two resistance values.

8.5   THE PRACTICAL AMPLIFIER

The ideal operational amplifier so far considered cannot be used to evaluate the input and output characteristics of actual amplifiers because these depend on the non-ideal properties of practical designs. Both the zeros and infinities we have so far assumed must, in reality, be finite quantities. We now briefly discuss the general electrical specifications of operational amplifiers.

8.5.1   Rated output

The operational amplifier clearly must have certain limitations on the available output signal swing. These limitations are specified by stating the rated output voltage v_0, and current i_0 for which linear operation applies. The limit on the output voltage swing is the saturation voltage which is normally a volt or so in excess of the rated output voltage. Although no damage will occur when the output is driven into saturation (as indeed it may in many non-linear applications), the overload recovery time may vary considerably from one particular application to the next. This is plainly of importance where high speed switching is concerned.

A typical amplifier transfer characteristic relating input to output voltage is shown in Fig. 8.17. Note that the gradient of this characteristic is equal to the amplifier gain A_v. Since the amplifier operates on finite values of power supply voltage, the characteristic exhibits a saturation effect slightly below the supply voltage.

8.5.2   Finite gain

Up to this point we have assumed the open-loop gain of the amplifier,

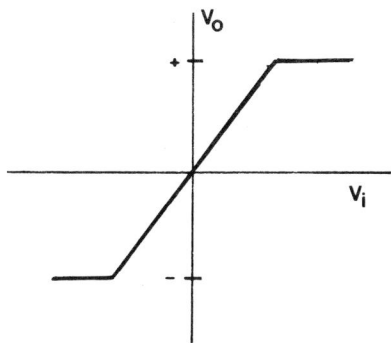

Figure 8.17

A_v, to be infinite. In the practical amplifier, it usually lies in the range 10^4 to 10^7. The effects of a finite open-loop gain can be calculated from the diagrams of Fig. 8.18(a) and (b).

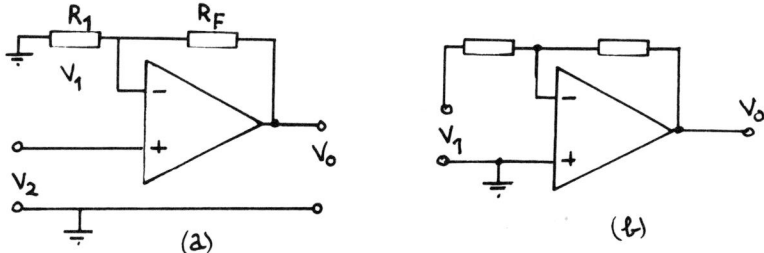

Figure 8.18

For the non-inverting amplifier, diagram (a), we have

$$v_o = A_v(v_2 - v_1) = A_v v_2 - \frac{v_o R_1}{R_1 + R_F}$$

$$A_v v_2 = v_o \left(1 + A_v \frac{R_1}{R_1 + R_F}\right)$$

The gain with feedback

$$A'_v = \cfrac{A_v}{1 + A \cfrac{R_1}{R_1 + R_F}} = \cfrac{\cfrac{R_1 + R_F}{R_1}}{1 + 1/\beta A_v} \qquad (8.5)$$

where

$$\beta = \cfrac{R_1}{R_1 + R_F}$$

For the inverting amplifier, diagram (b), we have

$$\frac{v_1 - v_s}{R_1} = \frac{v_s - v_o}{R_F} \qquad \text{and} \qquad v_o = -A_v v_s$$

Then

$$\frac{v_1}{R_1} = -\frac{v_o}{A_v R_1} - \frac{v_o}{A_v R_F} = -\frac{v_o}{R_F} \frac{R_F}{A_v R_1} + \frac{1}{A_v} + 1$$

$$\frac{v_o}{v_1} = \cfrac{-R_F/R_1}{1 + \left[A_v \left[\cfrac{R_1 + R_F}{R_1} \right] \right]}$$

Therefore

$$A'_v = \cfrac{-R_F/R_1}{1 + \cfrac{1}{\beta A_v}} \qquad (8.6)$$

Inspection of equations (8.5) and (8.6) show that the idealised gain equations derived earlier, that is

$$\frac{v_o}{v_1} = -\frac{R_F}{R_1} \qquad \text{for the inverting amplifier}$$

and

$$\frac{v_o}{v_2} = \frac{R_1 + R_F}{R_1} \qquad \text{for the non-inverting amplifier}$$

are modified in each case by what we might call a gain error term

$1/\beta A_V$. The closed loop gain of either amplifier may then be expressed as

$$\text{Actual closed-loop gain} \; \frac{\text{Idealised closed-loop gain}}{1 + 1/\beta A_V}$$

In the non-inverting circuit the closed loop gain is approximately $1/\beta$; in the inverting circuit it is approximately $(1 - 1/\beta)$.

Example 8.6   The ideal voltage follower has an infinite input resistance, zero output resistance and unity gain. Assess the input and output resistance of a practical operational amplifier having $R_i = 150$ kΩ, $R_o = 100$ Ω and $A_V = 10^4$ when used as a voltage follower.

The idealised model of Fig. 8.10 must be modified to take account of the finite values of R_i and R_o found in the practical amplifier. Fig. 8.19 represents such a model in diagram (a) with the equivalent circuit at (b). A load R_L is connected to the output terminals.

It will be of interest to derive the necessary equations as we go. For the actual input resistance, applying Kirchoff we have

$$v_1 = i_1 R_i + R_o(i_1 - i_2) + A_V(v_1 - v_o) \tag{i}$$

$$v_1 = i_1 R_i + R_L i_2 \tag{ii}$$

$$v_o = i_2 R_L \tag{iii}$$

Writing (ii) as

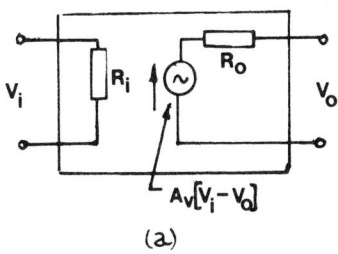

(a)

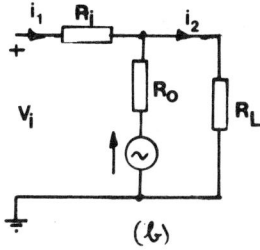

(b)

Figure 8.19

279

$$i_2 = \frac{(v_1 - i_1 R_i)}{R_L}$$

and substituting this into (i) along with equation (iii) yields

$$v_1 = i_1 R_i + R_o i_1 - R_o \frac{v_1 - i_1 R_i}{R_L} + A_v v_1 - A_v R_L \frac{v_1 - i_1 R_i}{R_L}$$

The input resistance with feedback is then found by rearrangement:

$$R_i' = \frac{v_1}{i_1} = \frac{R_i + R_o + R_o R_i / R_L + A_v R_i}{1 + R_o / R_L}$$

and multiplying top and bottom by R_L

$$R_i' = \frac{R_i R_L + R_o R_L + R_o R_i + A_v R_i R_L}{R_L + R_o}$$

For $A_v \gg 1$ and $R_L \gg R_o$ this reduces to

$$R_i' = \frac{(A_v + 1) R_L R_i}{R_L} + \frac{R_o (R_L + R_i)}{R_L}$$

$$\simeq A_v R_i$$

This accords with what we have deduced earlier about the effect of feedback on input resistance.

With $R_i = 150 \text{ k}\Omega$ without feedback

$$R_i' = 10^4 \times 150 \times 10^3 \ \Omega$$

$$= 1500 \text{ M}\Omega$$

For the output resistance, we first disconnect the load so making $i_2 = 0$. Then $v_{oc} = v_1 - i_1 R_i$

$$= v_1 - R_i \frac{v_1 (1 - A_v) + A_v v_{oc}}{R_i + R_o}$$

and solving for v_{oc} yields

$$v_{oc} = v_1 \frac{R_i + R_o + R_i(A_v - 1)}{R_i + R_o + A_v R_i}$$

$$= v_1 \frac{R_o + A_v R_i}{R_o + R_i(A_v + 1)}$$

Now shorting out the output terminals so that $v_o = 0$ we have

$$i_{sc} = \frac{v_1}{R_i} + \frac{A_v v_1}{R_o} = v_1 \cdot \frac{R_o + A_v v_1}{R_i R_o}$$

For $A_v \gg 1$ and $R_i \gg R$ we have for the feedback output resistance

$$R'_o = \frac{v_{oc}}{i_{sc}} = \frac{R_i R_o}{R_o + R_i(A_v + 1)} \simeq \frac{R_o}{A_v}$$

With $R_o = 100 \ \Omega$ and $A_v = 10^4$

$$R'_o = \frac{100}{10^4} = 0.01 \ \Omega$$

8.5.3   Unity gain-bandwidth

The open-loop gain of an operational amplifier varies with frequency
and like those amplifiers already discussed is conveniently represented as a Bode plot. The absolute value of voltage gain is plotted
in dB versus a decade logarithmic frequency scale. The shape of the
Bode plot is so similar for most operational amplifiers in common
usage that any open-loop plot may be approximated very closely from
the two basic specifications: one, the loop gain A_o and two, the
transition frequency. Fig. 8.20 illustrates a typical Bode plot for
an operational amplifier where the design is phase compensated to
restrict the slope of the roll off to the first-order 6 dB/octave.
This then ensures circuit stability at any closed loop gain.

The gain at any frequency relative to A_o is, of course, given by

$$A(\omega) = \frac{A_o}{1 + j\frac{\omega}{\omega_2}}$$

where ω_2 is the upper 3 dB frequency. If this expression is substi-

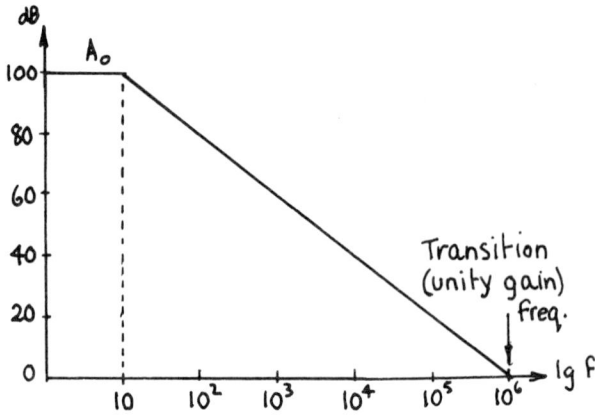

Figure 8.20

tuted into the gain equation for the non-inverting amplifier of Fig.
8.18(a), the closed loop gain A'(ω) becomes

$$\frac{\dfrac{R_1 + R_F}{R_1}}{1 + j\dfrac{\omega}{\omega_2} \cdot \left[\dfrac{R_1 + R_F}{R_1}\right]}$$

A plot of this equation, Fig. 8.21, shows that at low frequencies
where the loop gain is high, the closed loop gain is determined by
the feedback network. At high frequencies where the loop gain is
small, the open loop gain is the determining factor and the closed
loop gain merges with the open loop slope. We have already discussed
the constancy of the gain-bandwidth product.

We can now amplify the previous discussion on the stability of a
multistage amplifier. The rate of closure between open loop gain
and 1/β is a measure of the stability of the closed loop circuit. A
6 dB/octave closure rate is, as we have already seen, equivalent to
an open loop phase shift of 90°. A 12 dB/octave rate is then equivalent to a 180° open loop phase shift. If the loop phase shift is
-180° or greater where the loop gain is unity, the circuit will be
unstable when the feedback is applied. A rate of 18 dB/octave
would certainly lead to instability.

282

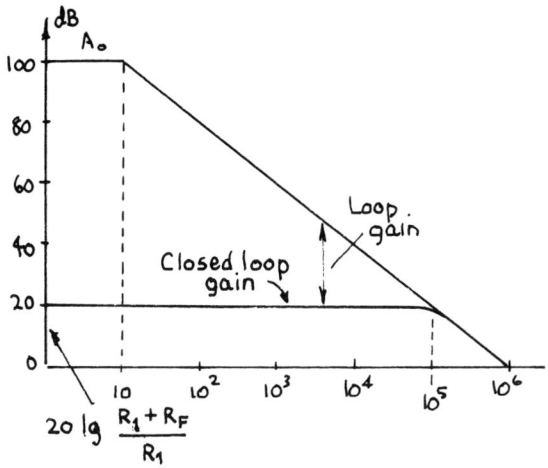

Figure 8.21

8.5.4   Slew rate and settling time

The nominal gain roll off of an operational amplifier is, as we have
seen, 6 dB/octave. The step function response of such an amplifier
is an exponential rising with a time constant of $1/\omega_2$ where ω_2 is the
closed loop bandwidth. Fig. 8.22 illustrates this. This performance, however, applies only for small signal levels so that the
amplifier is operating linearly. For large signals the amplifier
rise time is limited by what is known as the 'slew rate' which is a
measure of the maximum possible time rate of change of output voltage of which the amplifier is capable. Slew rate is usually measured
in volts per microsecond. Hence the maximum frequency at which an
amplifier can usefully operate is not only a function of the band-

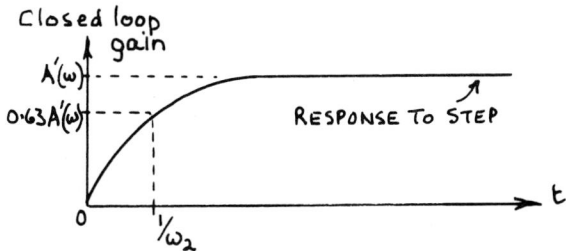

Figure 8.22

width but also of the slew rate. However, with a sinusoidal input waveform, whereas the bandwidth limits the gain at the higher frequencies, the slew rate leads to distortion. When the frequency and amplitude of an output sinusoid are such that the maximum rate of change exceeds the slew rate, distortion must follow; and the frequency at which the zero crossing gradient of a sine wave equals the slew rate is the maximum frequency for the rated output or full power response. Mathematically we may say, let the waveform be $v = \hat{V}.\sin \omega t$, then the time rate of change of v is $dv/dt = \omega\hat{V} \cos \omega t$ and this has a maximum value when $t = 0$, i.e. the zero crossing points on the time axis. Hence

$$\frac{dv}{dt} (max) = \omega\hat{V}$$

and for the amplifier to operate without distortion, the slew rate must exceed this value.

Usually the slew rate limit occurs in the first stages of an amplifier and is governed by the size of the capacitances associated with this stage. For any given amplifier the slew rate is constant so the designer can either go for a large output or a good high frequency response, but he cannot have both at once. The popular 741 integrated operational amplifier has a slew rate of about 0.5 V/µs, although the 741 S version can manage a slew rate of 20 V/µs.

In circuits where transients are the order of the day, such as D to A or A to D converters, the desired amplifier characteristic is a fast 'settling time' or recovery time. This is a parameter which is affected by the slew rate, the small signal frequency response, and external circuit parameters such as capacitive loading. It does not follow, however, that a fast slew rate and wide band width will automatically provide a fast recovery time.

Example 8.7   The 741 has a slew rate of 0.5 V/µs. What will be the maximum frequency at which the 741 will operate without distortion if an output amplitude of 1 V is provided?

The slew rate must equal (or exceed) the value of $\omega\hat{V}$, hence

$$f = \frac{slew\ rate}{2\pi V} = \frac{0.5}{2\pi \times 1 \times 10^{-6}}\ Hz$$

$$= 3.14\ MHz$$

It has been assumed that the 1 V amplitude referred to the 'peak' value. If the r.m.s. value is implied, the solution obtained must be divided by $\sqrt{2}$; the maximum frequency in that case will be 2.22 MHz.

Applications and wave shaping properties of operational and other amplifiers will be discussed in the following chapter.

8.6   A LABORATORY EXPERIMENT

Set up the circuit of Fig. 8.23, using a 741 operational amplifier I.C. This is the inverting configuration. The supply voltages can be obtained conveniently from two 9 V batteries.

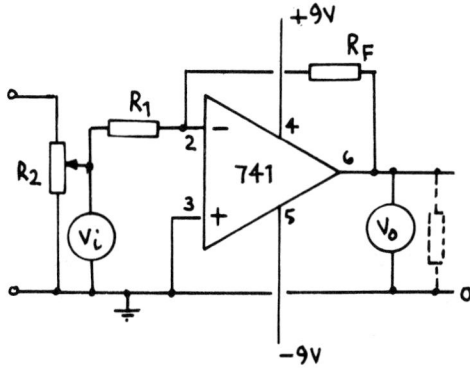

Figure 8.23

By changing the ratio R_F/R_1 the gain of the amplifier is varied; if v_1 and v_o are measured at each selected ratio, the actual closed loop gain can be calculated. Start with R_1 = 10 kΩ, R_F = 1 MΩ, so that A_V = -100. Use potentiometer R_2 to apply SMALL steady (d.c) voltages increments (both positive and negative) to the amplifier and record v_i and v_o for each setting of v_i. Tabulate the results and plot a graph of the MEASURED values of v_o against the CALCULATED values. Your graph should be of the form shown in Fig. 8.17 where the horizontal axis is v_o (calculated) = $v_i \times A_V$.

Slew Rate. Now put R_1 = R_F = 100 kΩ so that the gain is unity. Replace the voltmeter v_o with an a.c. millivoltmeter and shunt v_o with a 2.2 kΩ load resistor. Apply a sinusoidal signal of frequency about 1 kHz and amplitude 0.5 V r.m.s. to the input of the amplifier. Measure v_o as the input frequency is now increased and find the HIGHEST frequency at which the gain remains at unity. Repeat this procedure up to an input amplitude of about 5 V r.m.s. Where the gain deviates from unity, slew rate distortion has set in. (An oscilloscope will also show the onset of distortion though it is not so sensitive as the millivoltmeter monitoring).

Since slew rate $S = 2\pi f\hat{V}$, plot $1/\hat{V}$ (horizontally) against frequency f, and hence obtain a figure for S.

PROBLEMS 8

1. Make a list of the ideal properties of an operational amplifier.

2. When a basic analysis is made of an operational amplifier with feedback, what values are assigned for (a) the input current, (b) the input voltage? Explain.

3. How can an operational amplifier be used as an analogue adder?

4. A voltage follower has unity gain and is non-inverting. What is the purpose of such an amplifier? Explain briefly how it works.

5. What is the difference between a difference and a common-mode signal?

6. What is a differential amplifier? Name its advantages.

7. List the practical limitations on operational amplifier performance.

8. Why is the high frequency gain intentionally reduced in an operational amplifier?

9. Define the term 'slew rate'.

10. An operational amplifier has a slew rate of 1 V/μs when operating with a V_{cc} supply of ±9 V. What will be the switching time of this amplifier? Is this slew rate good or poor?

11. Figure 8.24 shows an inverting circuit where the amplifier can be considered to have a very high gain and input resistance. If $R_1 = R_F$, show that $V_o/V_i \approx -1$.

12. Three voltages V_1, V_2 and V_3 are to be added by using an opera-

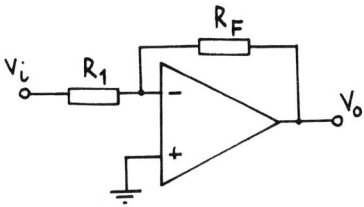

Figure 8.24

tional amplifier which may be considered to have ideal characteristics. The input resistors R_1, R_2 and R_3 (see Fig. 8.12) are each of value 100 kΩ. What must be the value of R_F such that $V_o = V_1 + V_2 + V_3$?

If the actual gain of the amplifier is -500, what must be the value of R_F if the output voltage is to be unchanged?

13. An operational amplifier has $A = 10^4$, $R_i = 10$ kΩ and $R_o = 1$ kΩ. If it is used as a unity gain buffer, show that the input and output resistances become 100 MΩ and 100 mΩ respectively.

14. What will be the output voltage of the circuit shown in Fig. 8.25?

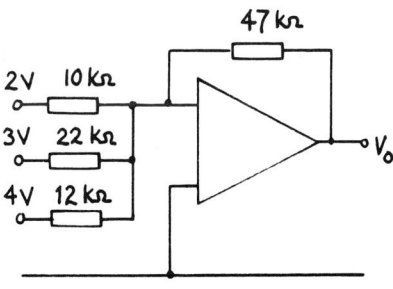

Figure 8.25

15. When wiring up a voltage follower, an inexperienced experimenter crossed over the + and - inputs to the amplifier so that V_i was connected to the - input and V_o was fed back to the + input. Derive an expression for the gain of his system and obtain a value for the input resistance. Is this circuit of no value or could an application be found for it?

287

16. An operational amplifier has a voltage gain of 100. Calculate the required values of R_F for an inverting gain (Fig. 8.11) and a non-inverting gain (Fig. 8.14) respectively, given that R_1 = 470 kΩ.

17. Draw a circuit diagram of an emitter-coupled differential amplifier. Draw an equivalent circuit and from this derive an expression for the signal output voltage. Show that for equal input signals the output voltage is zero. State the assumptions you have made in the derivation.

18. Briefly discuss the main causes of drift in transistor amplifiers, and explain why it is particularly troublesome in direct-coupled amplifiers.

In the emitter-coupled amplifier of Fig. 8.26, the input residual drift voltages are δv_1 and δv_2 for the transistors T_1 and T_2 respectively. Derive an expression for the output voltage v_o in terms of δv_1 and δv_2 and the circuit parameters, and hence show that for good common-mode rejection the emitter resistor R_E must be large enough to satisfy the inequality

$$R_E \gg \frac{h_{ie}}{2(h_{fe} + 1)}$$

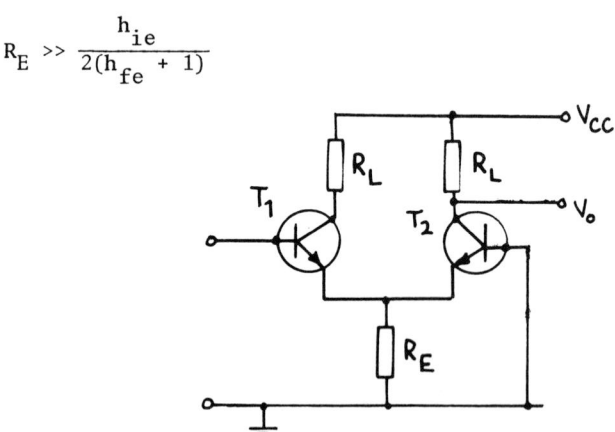

Figure 8.26

19. An operational amplifier has the following specifications:

    Nominal gain = 90 dB + 3 dB - 5 dB
    Input resistance 100 kΩ

The frequency response of this amplifier is shown in Fig. 8.27. If this amplifier is used with negative feedback in series with the input to give a nominal gain of 1000, obtain figures for (a) the possible gain spread, (b) the input resistance, (c) the bandwidth.

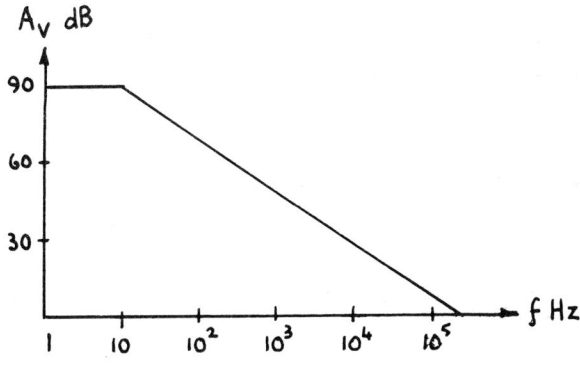

Figure 8.27

9 Oscillators and waveform generators

9.1   INTRODUCTION

The expression derived for the gain of an amplifier with feedback

$$A' = \frac{A}{1 - \beta A}$$

suggests that an increase in gain can be obtained if the denominator
$(1 - \beta A)$ is less than unity. However, this method of increasing gain
is rarely if ever used in amplifier systems because of the inherent
tendency that such amplifiers would have towards instability. In the
early days of radio, the gain of simple one- and two-valve receivers
(which at best were not particularly sensitive) was boosted by a controlled amount of such positive feedback which went under the general
name of 'reaction', and with careful adjustment on the part of the
operator, such a control did indeed produce an appreciable increase
in the gain. If the feedback was increased, however, and βA became
equal to unity, then theoretically the gain became infinite and the
programme was lost to the accompaniment of howls and whistles from
the headphones or loudspeaker and suitable expletives from the operator. The amplifier had, in fact, now become an 'oscillator'.

Oscillators are of great importance in electronic and communica-
tion engineering for they provide (intentional) signal sources for
electronic measurements and are generally employed as such in combination with other measuring devices under the name of signal
generators or function generators. Oscillators operating at fixed
frequencies or operating over restricted ranges of frequency are
found in transmitters and in all radio and television receivers.

An oscillator is essentially a power converter in the sense that its only input is the d.c. power supply and its output is a continuous waveform which may or may not be sinusoidal in shape. With positive feedback, that part of the output fed back to the input provides the amplifier with its own input. In a circuit diagram of an oscillator, therefore, we do not expect to find a terminal which is specifically the input signal terminal, although there will be an output terminal from which the generated waveform may be taken off.

9.2   CONDITIONS FOR OSCILLATION

The statement $\beta A = 1$ implies that the loop gain of the system is unity and that zero phase shift (or multiples of 2π) occurs around the loop. The feedback voltage is then identical to the input voltage and the amplifier is able to supply its own input signal. In making $\beta A = 1$ there are two implied conditions for the circuit to be oscillatory which can also be expressed in the form $1/0^{\circ}$, and from this the two conditions are (a) the magnitude of the gain round the loop shall be unity to maintain the proper level of input signal, (b) the phase shift round the loop shall be zero or $2\pi n$, where n is an integer, or 180° different from the condition normally required for negative feedback.

In addition the amplifier must be capable of supplying sufficient power to compensate for that which is dissipated in the resistance of the feedback network and still have enough available for its own input requirements and any necessary external loading.

A consideration of these requirements leads to the conclusion that the 'amplifier' circuit forms that part of the feedback loop which maintains the 'amplitude' of the output, and that the 'β-network' determines the 'frequency' of the output. In practice, the loop gain must be slightly greater than unity, for if it were precisely equal to 1, then variations in the supply voltages, ageing of components and possibly temperature and other environmental changes would in all likelihood result in a collapse of the oscillations because of a re-duction (even though temporary) in the overall gain. When the loop gain is made greater than unity, more signal is fed back than is actually required for oscillation to occur and a build-up in signal amplitude around the loop quickly follows; this build-up cannot continue indefinitely but is rapidly limited by non-linearities within the amplifier and by the finite value of the supply voltage.

9.3   CLASSES OF OSCILLATOR

There are two main classes of positive feedback systems: (a) those in which the generated waveform is sinusoidal; these are known as sinusoidal or harmonic oscillators; (b) those in which the generated waveform is markedly non-sinusoidal, being characterised by sudden transitions from one condition of stability to the next. These systems come under the general heading of 'relaxation' oscillators. Whatever the shape of the output wave, it always has some definite frequency or repetition rate and this is determined by either an inductance-capacitance or a resistance-capacitance network in the feedback loop.

9.4   SINUSOIDAL OSCILLATORS

Two performance features need to be stable and well designed: (a) frequency, (b) the generated amplitude. These parameters are usually linked to the conditions mentioned above for oscillations to be maintained, thus the most usual circuit arrangement is as shown in Fig. 9.1.

At the desired frequency only, $\underline{/\beta A} = 0^{\circ}$. This condition may be achieved by any arrangement where the phase shift in the amplifier A is equal in magnitude but opposite in sign to the phase shift in β. However, an amplifier with an even or odd number of stages will have a phase shift between input and output of either 0° or 180° and this may be considered constant providing that the frequency is within the normal mid-frequency passband of the amplifier's coverage. It is possible to have other phase shifts at the very high or very low frequencies which can also produce 'equivalent' phase shifts to those at the mid-frequencies but while these can prove troublesome in the design of stable amplifiers, they are less important in the design of

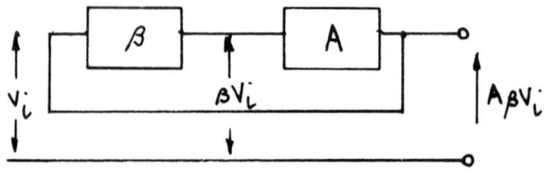

Figure 9.1

292

oscillators.

So, for 180° phase shift in A, the β network must provide a further shift of 180°; and for zero shift in A, the network must provide a further 360° (or 0°) shift. This condition will be met at some frequency ω_1 in the β network phase characteristic; thus, as mentioned above, it is the β network and not the amplifier which determines the frequency of oscillation to be at or very close to ω_1. At this frequency, the magnitude of the loop gain should be >1 for small signal levels and should become exactly equal to 1 for the desired output amplitude. Thus it is a feature of the amplifier and not the β network which maintains the oscillation amplitude.

The two requirements can be designed for separately. Because the phase shift around the loop must be zero, such oscillators are called 'phase-shift oscillators'; in a sense this term actually covers all types of oscillator except those using the negative resistance characterstics of a circuit, though the name 'phase shift oscillator' is usually reserved for oscillators employing a passive phase shifting network in the feedback path.

9.4.1 Phase shift networks

Phase shift networks may be constructed using resistors, capacitors and inductors. Resistance-capacitance networks will be dealt with first, and the basic circuits are shown in Fig. 9.2(a) and (b). In these circuits as we have seen in the chapter on small-signal amplifiers, the limit of phase change is 90° but there is a disadvantage in the severe drop in V_o as the angle approaches 90°. A series of such networks is therefore necessary to secure the desired phase

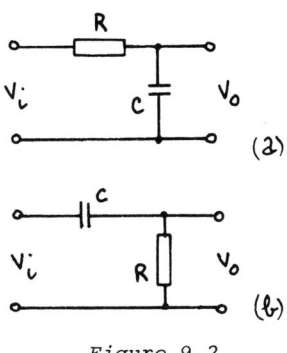

Figure 9.2

change of 180° without excessive attenuation of the input signal.

Consider three identical phase retard networks connected in series and let the input current be i_1, then allowing circulating currents to flow in the various loops as shown in Fig. 9.3 we have a current transfer ladder network where the current gain is given by i_0/i_1. We require to find this gain and the frequency at which the overall

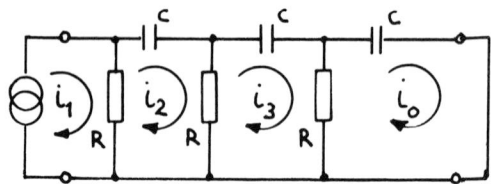

Figure 9.3

phase shift is 180°. Let the reactance of $C = jX$, and then

$$i_1 R = i_2(2R + jX) - i_3 R \tag{i}$$

$$i_2 R = i_3(2R + jX) - i_0 R \tag{ii}$$

$$i_3 R = i_0(R + jX) \tag{iii}$$

Substituting for i_2 from (ii) into (i)

$$i_1 R = (2R + jX)^{i_3}/R - (2R + jX)i_0 - i_3 R \tag{iv}$$

Substituting for i_3 from (iii) into (iv)

$$i_1 R = (2R + jX)^2(R + jX)\frac{i_0}{R} - (2R + jX)i_0 - (R + jX)i_0$$

Hence

$$i_1 R^3 = (R^3 - 5RX^2 + j6XR^2 - jX^3)i_0 \tag{v}$$

and the current gain

$$\frac{i_0}{i_1} = \frac{R}{R^3 - 5RX^2 + j6XR^2 - jX^3}$$

For zero and 180° phase shift between input and output the j-terms are zero:

$$6XR^2 - X^3 = 0$$

$$X^3 = 6R^2X$$

Therefore

$$X = 0 \quad \text{or} \quad X = \sqrt{6}.R = {}^1/\omega C$$

Therefore

$$\omega = \frac{1}{\sqrt{6}.CR} \quad \text{or} \quad f = \frac{1}{2\pi CR\sqrt{6}} \tag{9.1}$$

This is the frequency at which 180° phase shift occurs. The solution $X = 0$ would represent zero phase shift in a ladder in which the capacitors were absent and i_o would be equal to i_1; this case is of no interest. At the frequency given in (9.1) above, the current gain becomes

$$\frac{i_o}{i_1} = \frac{R^3}{R^3 - 5RX^2}$$

But $X^2 = 6R^2$, hence

$$\frac{i_o}{i_1} = \frac{R^3}{R^3 - 30R^3} = \frac{1}{29}$$

When used in a feedback network, therefore, this ladder will make $\beta = -1/29$. Since the condition for oscillation is that $\beta A \geq 1$, the current gain of the amplifier must be at least -29 to compensate for the attenuation in the β-network.

The ladder discussed was a phase advance network; if a phase retard is considered, C and R are interchanged and the analysis (which you might care to try) leads to a frequency for 180° phase shift of

$$f = \frac{\sqrt{6}}{2\pi CR} \tag{9.2}$$

Clearly, the ladder discussed must be driven from a high impedance source and must feed into a low impedance load, and a common-emitter amplifier satisfies both these conditions reasonably well. Conversely, the equivalent FET circuit would use a voltage transfer ladder (phase retard) which would require a low impedance source and a high impedance load. Figure 9.4 shows a practical form of the

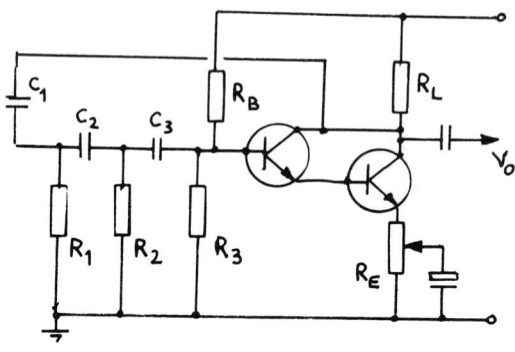

Figure 9.4

phase shift oscillator using bipolar transistors. The amplifier takes the form of a Darlington pair; this provides adequate current gain and at the same time provides the required impedance matching to the input and output of the ladder network. Some degree of negative feedback is provided by the unbypassed part of the emitter resistance and such a form of control is usually necessary with this form of oscillator. Distortion of the output waveform (which should be sinusoidal) can be very serious in phase shift systems of this sort because in the absence of some sort of gain control, the only factor available to limit the output amplitude is the non-linearity in the transistors themselves at high signal excursions. Further, the variation of phase shift with frequency in the ladder on either side of the 180° reference is not sharp, hence considerable gain round the loop still exists at harmonics of the main frequency and distortion products are amplified along with the wanted output. From this point of view the phase retard network (which is a low-pass filter) has an advantage when used in the feedback path in that it attenuates the higher harmonics of the fundamental frequency.

A disadvantage of the phase-shift oscillator is that it does not lend itself easily to being tuned over a range of frequencies. Either the three resistors or the three capacitors must be adjustable simultaneously. For this reason, this kind of oscillator is usually designed for fixed frequency operation.

<u>Example 9.1</u>   A phase shift oscillator uses a single transistor having h_{fe} = 250 in the amplifier section of the feedback loop, see Fig. 9.5. and generates a frequency of 500 Hz. If the component values are as indicated on the diagram, calculate the input resistance of the transistor.

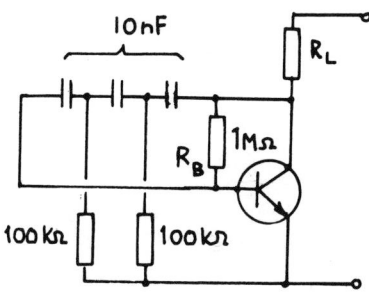

Figure 9.5

There appear to be only two similar sections in the feedback network, so the third section must be provided by the 'end' C and the input resistance of the transistor.   Now the phase shift provided by each of the two similar RC sections is given by

$$\phi = \tan^{-1} \frac{1}{\omega CR}$$

$$= \tan^{-1} \frac{1}{2\pi \times 500 \times 10^{-8} \times 10^{4}}$$

$$= 72.6^{\circ}$$

The total shift through the two stages is therefore 145.2°. The remaining shift contributed by C and the input resistance of the transistor is therefore $(180^{\circ} - 145.2^{\circ}) = 34.8^{\circ}$. Hence

$$R_{i} = \frac{1}{2\pi f \times C \times \tan 38.4^{\circ}}$$

$$= \frac{10^{8}}{2\pi \times 500 \times 0.695} \; \Omega = 45.8 \text{ k}\Omega$$

This resistance includes the shunting effect of the bias resistor R_B. The effective input resistance of the transistor is therefore about 50.7 kΩ.

This example illustrates the fact that the three RC networks in the ladder do not have to be identical and indeed they rarely are. The theoretical considerations discussed above have assumed zero and infinite impedances at the ports of the ladders and in practical circuits this cannot be so. This also means that the minimum gain figure of -29 is not likely to provide sufficient feedback to keep the system in oscillation. The circuit of Fig. 9.6(a) shows the equivalent a.c circuit using h-parameters. This can be simplified

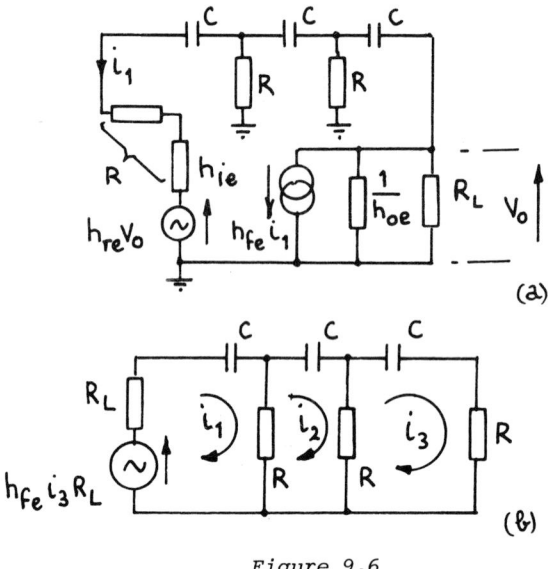

(a)

(b)

Figure 9.6

by ignoring h_{re} as being very small and $1/h_{oe}$ as very large, so that the simplified circuit of Fig. 9.5(b) results. Here the source has been transformed into a voltage generator. From the loop currents we now have

$$i_1(R_L + R + jX) - i_2R + i_3h_{fe}R_L = 0$$

$$i_1R - i_2(2R - jX) - i_3R = 0$$

$$i_2R - i_3(2R - jX) = 0$$

From these equations, working as before, we find that the frequency at which the phase shift is 180° is

$$\omega = \frac{1}{\sqrt{C}(4RR_L + 6R^2)}$$

and substituting this back into the real part of the solution gives the minimum device gain as

$$h_{fe} = 23 + \frac{29R}{R_L} + \frac{4R_L}{R}$$

9.4.2   The Wien network

The Wien network which also uses only resistors and capacitors is shown in Fig. 9.7. This circuit will obviously not behave in the same way as the simple ladder sections already discussed. From

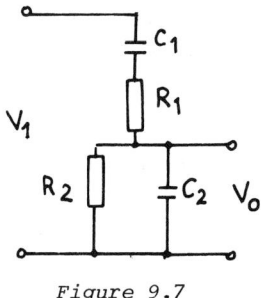

Figure 9.7

the figure

$$Z_1 = R_1 - jX_1 \quad \text{and} \quad Z_2 = \frac{-jX_2R_2}{R_2 - jX_2}$$

Then

$$\frac{v_1}{v_o} = \frac{Z_1 + Z_2}{Z_2} = \frac{R_1 - jX_1 + \dfrac{-jX_2R_2}{R_2 - jX_2}}{\dfrac{-jX_2R_2}{R_2 - jX_2}}$$

$$= \frac{R_1 R_2 - X_1 X_2 - j X_1 R_2 - j X_2 R_1 - j X_2 R_2}{-j X_2 R_2}$$

$$= j\frac{R_1}{X_2} - j\frac{X_1}{R_2} + \frac{X_1}{X_2} + \frac{R_1}{R_2} + 1$$

For zero phase shift the j-terms must be zero:

Therefore

$$\frac{R_1}{X_2} = \frac{X_1}{R_2}$$

Hence

$$R_1 R_2 = X_1 X_2 = \frac{1}{\omega^2 C_1 C_2}$$

Therefore

$$\omega = \frac{1}{\sqrt{R_1 R_2 C_1 C_2}} \quad \text{or} \quad f = \frac{1}{2\pi\sqrt{R_1 R_2 C_1 C_2}}$$

This is the frequency at which the phase shift is zero; with this condition the voltage gain of the network is given by the real part of the expression, so

$$\frac{v_1}{v_0} = \frac{X_1}{X_2} + \frac{R_1}{R_2} + 1$$

When $R_1 = R_2 = R$ and $C_1 = C_2 = C$

$$\frac{v_1}{v_2} = 3 \quad \text{so that} \quad \beta = \frac{1}{3}$$

and the frequency of zero shift becomes $f = \frac{1}{2\pi CR}$

Since the criterion for oscillation is $\beta A \geqslant 1$, the voltage gain of the amplifier used with the Wien network must be 3 or greater and the amplifier must have zero phase shift. A single common-emitter

300

stage gives 180° phase shift, so two such stages are required. The
very high gain possible with such a circuit enables very heavy negative feedback to be used so that the gain is reduced to the required
value and the amplitude of oscillation is stable.

A Wien oscillator is shown in Fig. 9.8. This version uses two
bipolar transistors in a conventional two stage amplifier circuit

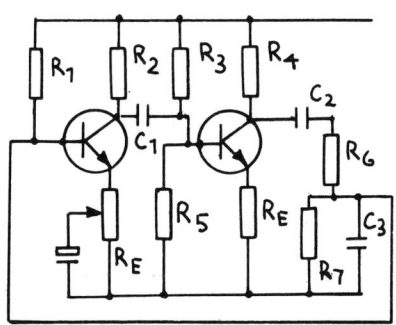

Figure 9.8

providing an overall phase shift of 360° (0°) and the Wien network is
connected across the output terminals. The voltage across the lower
section of the network provides the feedback signal. Notice again
how negative feedback is introduced by way of the unbypassed part of
the emitter load of transistor T_1; by çareful adjustment of this, a
very pure sinusoidal output waveform is obtained. There is one further point of interest: the lower resistor of the Wien network, R7,
forms in conjunction with R_1 the bias divider for transistor T_1.

The operational amplifier is well suited as the amplifier section
of a Wien oscillator, and Fig. 9.9 shows such a circuit. Here the
amplifier is used in the non-inverting mode which, assuming an infinite open-loop gain, gives the closed-loop gain as $1 + R_2/R_1$. If R_2
= $2R_1$, this gain will be 3 and oscillations will just be maintained.
R_1 is usually replaced by a non-linear device, such as a 6 V, 50 mA,
bulb; the loop-gain is then dependent upon the output amplitude, for
if this tends to increase, say, then the current through the bulb
will increase and its effective resistance will rise, so increasing
the amount of feedback and compensating for the initial build up in
amplitude.

The Wien oscillator has several advantages over the phase shift

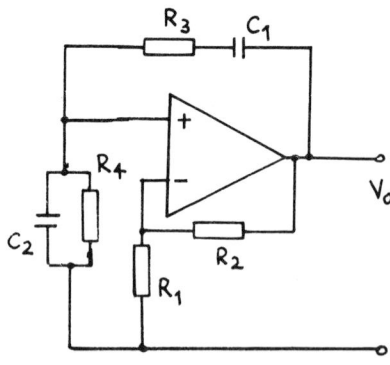

Figure 9.9

oscillator previously discussed. It is easily tuned over a very wide frequency range by using ganged capacitors or resistors (ganged resistors are usually chosen with the capacitors switched to provide frequency ranges), it operates down to very low frequencies with excellent waveform, and for a given variation in C it provides a much wider frequency variation than do circuits using inductance and capacitance, since f varies as $1/C$ and not $1/\sqrt{C}$ as in an oscillator using LC feedback. In fact, it is in the elimination of large inductances that the Wien oscillator is such a popular low frequency generator.

Example 9.2   Am amplifier with an overall phase shift of 0° is used with a phase shifting network as shown in Fig. 9.10. If the input resistance of the amplifier is very high, show that for the network

$$\frac{v_o}{v_1} = \frac{1}{3 + j(\omega CR - 1/\omega CR)}$$

and that the frequency of oscillation will be $\omega = 1/CR$.

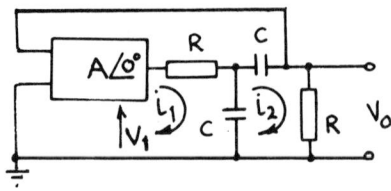

Figure 9.10

302

If in the actual amplifier R_i = 1 MΩ, R_o = 10 kΩ and the resistors are each 100 kΩ, what value capacitors will be required for a frequency of 200 Hz?

From the circuit as shown we have

$$v_1 = i_1 R + (i_1 - i_2) \frac{1}{j\omega C} \qquad\qquad (i)$$

$$v_1 = i_1 R + i_2 (R + \frac{1}{j\omega C}) \qquad\qquad (ii)$$

$$v_o = i_2 R$$

Subtracting (ii) from (i) yields

$$i_1 = i_2 \{\frac{2}{j\omega C} + R\} j\omega C$$

$$= \frac{v_o}{R} \{2 + j\omega CR\}$$

and substituting into (ii) gives

$$v_1 = \frac{v_o}{R} \{2 + j\omega CR\} R + \frac{v_o}{R} \{\frac{1}{j\omega C} + R\}$$

$$= v_o \{3 + j(\omega CR - \frac{1}{\omega CR})\}$$

The result now follows.

For zero phase shift in the network, the j-terms must be zero, hence

$$\omega CR = \frac{1}{\omega\, CR}$$

and

$$\omega = \frac{1}{CR}$$

Notice that the ratio v_o/v_1 now equals 1/3.

With resistors of value 100 kΩ, the frequency will be 200 Hz when

$$200 \times 2\pi = \frac{1}{10^5 \times C}$$

Therefore

$$C = \frac{10^6}{2\pi \times 200 \times 10^5} \ \mu F$$

$$= 0.008 \ \mu F.$$

Now because of the finite input and output resistances of the amplifier, the actual resistance values would have to be modified slightly; with R_o = 10 kΩ, this is in series with the left-hand re-sistor and if the total resistance is to be 100 kΩ, the component actually in the network will be 90 kΩ. Also R_i = 1 MΩ and this shunts the right-hand resistor; the total resistance is to be 100 kΩ so the component actually needed is 111 kΩ.

9.5   TUNED LC OSCILLATORS

At low frequencies the use of feedback networks containing inductors is prohibited by the bulkiness of the inductors and their inherent losses; in the same way, feedback networks containing resistors and capacitors become inconveniently small at high frequencies and it is then that inductors appear on the scene. Inductance-capacitance tuned circuits are used in selective amplifiers and in circuits where a range of frequencies, usually above some 100 kHz, have to be generated i.e. signal generators.

The basic tuned LC circuit is shown in Fig. 9.11. Here the source resistance is R_s and the loss resistance R_1. From the input

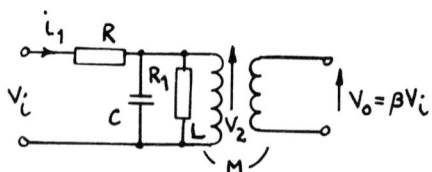

Figure 9.11

loop we have

$$v_1 = i_1 R_s + v_2 \tag{i}$$

$$i_1 = v_2 j\omega C + \frac{v_2}{R_1} + \frac{v_2}{j\omega L} \tag{ii}$$

With the output terminals open-circuited

$$v_o = j\omega M \frac{v_2}{j\omega L} = v_2 \frac{M}{L} \tag{iii}$$

i.e. $j\omega M$ times the current through L.

From these three equations i_1 and v_2 can be eliminated leaving one equation relating v_1 and v_o which is

$$\beta = \frac{v_o}{v_1} = \frac{M}{L} \frac{1}{\left[1 + \frac{R_s}{R_1} \right] + j \left[\omega C R_s - \frac{R_s}{\omega L} \right]} \tag{iv}$$

This equation will give the conditions for zero or 180° phase shift when the j-terms are zero, which is when either $R_s = 0$ or $\omega C - 1/\omega L = 0$. The condition $R_s = 0$ gives $\beta = M/L$ which is unselective to frequency and of no further interest. The condition $\omega = 1/\sqrt{LC}$ when substituted into (iv) gives the gain of the network as

$$\beta = \frac{M}{L} \frac{R_1}{R_s + R_1}$$

The actual phase change achieved, 0° or 180°, depends upon the connections to the coil; reversing one coil changes the sign of M, and hence the sign of β.

9.5.1   Stability of frequency

An amplifier has small but significant phase variations resulting from component value variations due to temperature or other environmental changes, for example, a capacitor may have a temperature coefficient of 200 parts per million which represents a frequency change of 0.2% for a 10° change in temperature. Such phase variations can be reduced (apart from a careful choice of circuit components) if the rate of change of phase with frequency in the feedback network is very high. This is particularly important in tuned systems. The slope of the phase-frequency character about a given frequency ω is indicated in Fig. 9.12; as the frequency defining agent, the β-network can be designed from high quality components to ensure that dφ/dω is large.

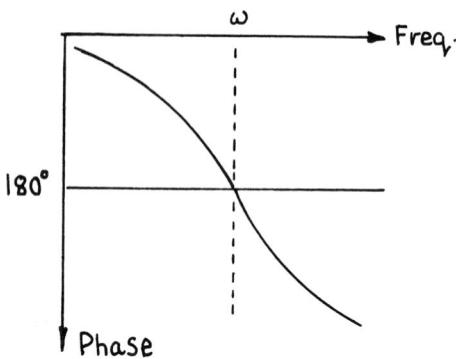

Figure 9.12

Returning to equation (iv) above which gives us the phase change conditions between v_1 and v_0 from the ratio of the imaginary and real parts, then when φ is close to 0^0 or 180^0 we can say

$$\phi = \tan \phi = \frac{- \{R_s \omega C - \frac{R_s}{\omega L}\}}{1 + \frac{R_s}{R_1}}$$

$$= \frac{- R_s R_1 (\omega C - 1/\omega L)}{R_s + R_1}$$

Differentiating we have

$$\frac{d\phi}{d\omega} = \frac{-R_s R_1}{R_s + R_1} \cdot \left[C + \frac{1}{\omega^2 L} \right]$$

At or near resonance $C = 1/\omega^2 L$, hence

$$\frac{d\phi}{d\omega} = - \frac{2}{\omega} \left[\frac{R_s \parallel R_1}{\omega L} \right] = - \frac{2Q}{\omega} \qquad (9.3)$$

Hence a necessary condition for phase stability at a given frequency, is that the circuit Q should be large. This obviously follows from earlier work on resonance. If we rewrite (9.3) as

$$\frac{d\omega}{\omega} = -\frac{d\phi}{2Q}$$

we obtain an expression for the fractional frequency change.

Notice that for a large Q-factor, the source resistance R_s must be large. This condition is met in both bipolar and field-effect transistors if the tuned circuit is wired into the colector or drain circuit.

9.5.2   The tuned collector oscillator

A large number of circuits use parallel (and sometimes series) tuned LC circuits in the feedback loop, and the tuned-collector, common-emitter oscillator is a useful introductory example. Fig. 9.13(a) shows a typical circuit using a bipolar transistor, though a FET may be equally employed. The circuit will oscillate when the transformer is phased to provide a 180° phase shift, and M is large enough.

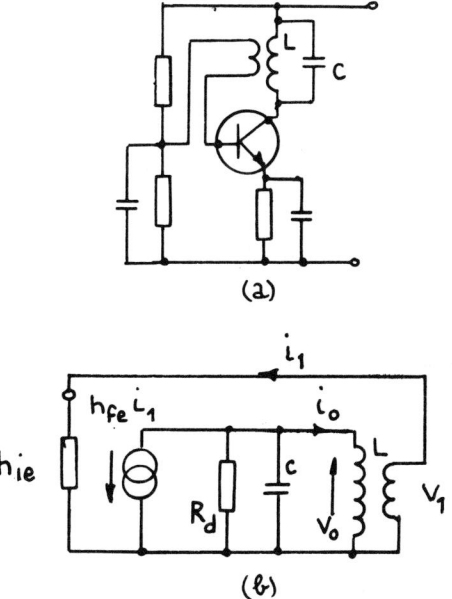

Figure 9.13

To analyse the circuit, we resort as usual to the equivalent circuit which is shown in Fig. 9.13(b). The tuned circuit consisting of L and C has been replaced by the parallel combination of L, C and the dynamic resistance R_d = L/CR. A portion of the output voltage developed across this circuit is then fed back as input to the amplifier. For simplification h_{re} and h_{oe} have been taken as negligible quantities. Now the load impedance

$$Z_L = \frac{1}{\dfrac{1}{R_d} + \dfrac{1}{j\omega L} + j\omega C}$$

Also

$$v_o = -h_{fe}i_1 Z_L$$

$$i_o = \frac{v_o}{j\omega L} = \frac{-h_{fe}i_1 Z_L}{j\omega L} \tag{i}$$

The induced voltage

$$v_1 = j\omega M i_1$$

Therefore

$$i_1 = \frac{j\omega M i_1}{h_{ie}}$$

$$= -\frac{h_{fe}}{h_{ie}} \cdot \frac{M}{L} \cdot i_1 Z_L \quad \text{from (i)}$$

Therefore

$$\frac{M}{L} = -\frac{h_{ie}}{h_{fe}} \cdot \frac{1}{Z_L}$$

$$= -\frac{h_{ie}}{h_{fe}} \left[\frac{1}{R_d} + \frac{1}{j\omega L} + j\omega C \right]$$

Equating the imaginary parts to zero:

$$\frac{1}{j\omega L} + j\omega C = 0$$

Therefore

$$\omega = \frac{1}{\sqrt{LC}}$$

Equating real parts

$$M = -L \frac{h_{ie}}{h_{fe}R_d} \tag{9.4}$$

These solutions give the frequency of oscillation and the least value of M necessary to maintain oscillation.

Example 9.3   A tuned collector oscillator has a tuned circuit made up of an inductance of 1 mH having a resistance of 4 Ω, in parallel with a lossfree capacitor of 500 pF. Calculate the frequency of oscillation and the least value of M necessary, given that the transistor has h_{ie} = 2 kΩ, h_{fe} = 50.

$$R_d = \frac{L}{CR} = \frac{10^{-3}}{5 \times 10^{-10} \times 4} \ \Omega = 500 \ k\Omega$$

$$M = -L \frac{h_{ie}}{h_{fe}} \cdot \frac{1}{R_d} = -\frac{10^{-3} \times 2 \times 10^3}{50 \times 500 \times 10^3} \ H$$

$$= -0.08 \ \mu H$$

The negative sign simply arises because of the assumed direction of connection of the coils.

The frequency follows from the usual expression for the high-Q parallel circuit

$$f = \frac{1}{2\pi\sqrt{LC}} = \frac{1}{2\pi\sqrt{10^{-3} \times 5 \times 10^{-10}}}$$

$$= \frac{10^7}{14\pi} \ Hz = 227.4 kHz$$

We can express equation (9.4) above in another way by noticing that, since R_d is the effective dynamic load on the transistor, $h_{ie}/(h_{fe}R_d)$ is equivalent to $1/A$, where A is the voltage gain of the transistor. Hence (9.4) may be expressed as M = L/A or for sustained oscillation

$$\frac{MA}{L} > 1$$

9.5.3 Colpitts oscillator

There are several forms of tuned oscillators which operates on the fact that there is zero phase change between the collector and emitter of a transistor when it is used in the common-base configuration. The Colpitts oscillator is one such example, and the basic circuit arrangement is shown in Fig. 9.14. Normal bias is applied to

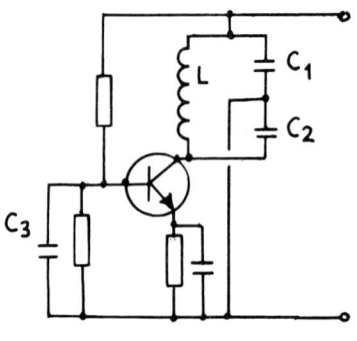

Figure 9.14

the base of the transistor which is a.c grounded by way of capacitor C_3 and inductor L is tuned by C_1 and C_2 effectively in series across L. The centre point of the capacitors is earthed, which makes frequency adjustment particularly easy by ganging the capacitors and earthing the common moving plates. Oscillators using a common-base amplifier have good high frequency performances. (And you should be able to say why.)

To analyse the circuit we need an equivalent model. We can get this by noticing that the tuned circuit impedances are $1/j\omega C_1$, $1/j\omega C_2$ and $j\omega L$. Let these be denoted by Z_1, Z_2 and Z_3 respectively and the circuit reduces to the form shown in Fig. 9.15(a); from this,

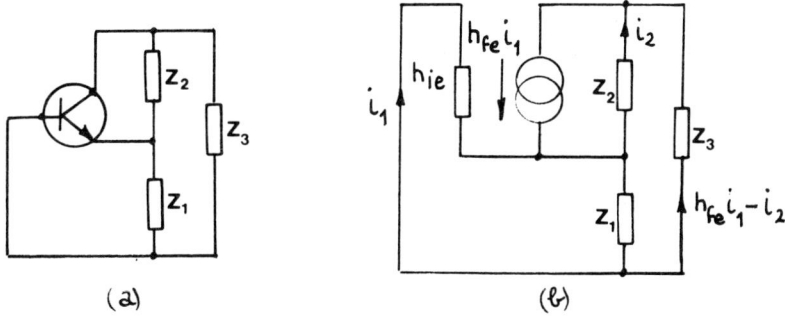

Figure 9.15

the h-parameter model of Fig. 9.15(b) is derived. From the input loop

$$i_1 h_{ie} + [(h_{fe} + 1)i_1 - i_2]Z_1 = 0$$

Therefore

$$i_1[h_{ie} + (h_{fe} + 1)Z_1] = i_2 Z_1 \qquad \text{(i)}$$

From the output loop

$$0 = i_2 Z_2 - (h_{fe}i_1 - i_2)Z_3 - [(h_{fe} + 1)i_1 - i_2]Z_1$$

Therefore

$$i_1[(h_{fe} + 1)Z_1 + h_{fe}Z_3] = i_2(Z_1 + Z_2 + Z_3) \qquad \text{(ii)}$$

Dividing (ii) by (i) gives us

$$\frac{(h_{fe} + 1)Z_1 + h_{fe}Z_3}{h_{ie} + (h_{fe} + 1)Z_1} = 1 + \frac{Z_2}{Z_1} + \frac{Z_3}{Z_1}$$

Hence

$$\frac{(h_{fe} + 1) + h_{fe}\dfrac{Z_3}{Z_1}}{\dfrac{h_{ie}}{Z_1} + (h_{fe} + 1)} = 1 + \frac{Z_2}{Z_3} + \frac{Z_3}{Z_1}$$

Substituting for Z_1, Z_2 and Z_3 yields

$$\frac{(h_{fe} + 1) - h_{fe}\omega^2 LC_1}{j\omega C_1 h_{fe} + (h_{fe} + 1)} = 1 + \frac{C_1}{C_2} - \omega^2 LC_1$$

Then

$$\omega^2 LC_1 + \frac{(h_{fe} + 1)C_1}{C_2} + j\omega C_1 h_{ie} \left[1 + \frac{C_1}{C_2} - \omega^2 LC_1 \right] = 0$$

Equating the imaginary terms to zero

$$1 + \frac{C_1}{C_2} - \omega^2 LC_1 = 0$$

Therefore

$$\omega^2 = \frac{1}{L}\left[\frac{1}{C_1} + \frac{1}{C_2} \right] \tag{iii}$$

from which

$$f = \frac{1}{2\pi} \sqrt{\frac{1}{L}\left[\frac{1}{C_1} + \frac{1}{C_2} \right]}$$

$$= \frac{1}{2\pi\sqrt{LC}}$$

where $1/C = 1/C_1 + 1/C_2$. This is a familiar expression except that the effective tuning capacitance is C_1 and C_2 in series. Equating the real terms to zero gives us

$$(h_{fe} + 1)\frac{C_1}{C_2} - \omega^2 LC_1 = 0$$

and substituting for ω^2 from (iii)

$$h_{fe} \frac{C_1}{C_2} + \frac{C_1}{C_2} - 1 - \frac{C_1}{C_2} = 0$$

or

$$h_{fe} = \frac{C_2}{C_1}$$

Hence for oscillations to be sustained

$$|h_{fe}| > \frac{C_2}{C_1}$$

Since C_1 is made equal to C_2 in most practical circuits, there is no problem in obtaining enough gain.

9.6   RELAXATION OSCILLATORS

The so-called family of relaxation oscillators generate non-sinusoidal outputs such as square or triangular waves, which may or may not be under the control of an external circuit. The basic circuit form of this type of oscillator is the free-running multivibrator or 'astable multivibrator' and several other useful forms are derived from this. We begin with a brief study of the transistor as a switching device.

9.6.1   The transistor switch

When analysing an active device such as a switch, the parameters of interest are the switching times and the ON and OFF resistances. A perfect switch would have zero ON resistance and an infinite OFF resistance and the time taken to switch from either of these states to the other would be zero. A mechanical switch comes very close to the ideal conditions for the ON and OFF resistance, but it is slow in its switching speed. Another drawback of mechanical switches is contact bounce; the contacting points oscillate over a very short period before becoming finally established. For the high speed operations required from switching systems in digital equipment and in oscillators generating pulse waveforms, the electronic switch is vastly superior to any form of mechanical switch, and both bipolar

313

and field-effect transistors find the widest employment in their
switching abilities.

Figure 9.16 shows a load line drawn across the output characteristics of either a bipolar or field-effect transistor. For linear
amplification that part of the load line lying between the points A

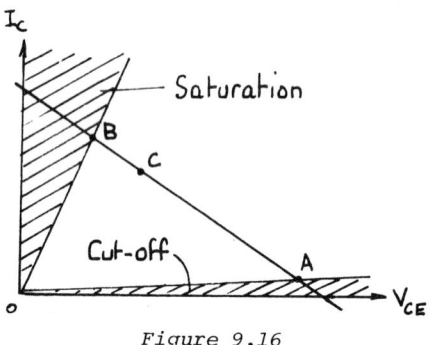

Figure 9.16

and B are used; for switching purposes the saturation and cut-off
regions are employed. Point A represents the upper limit of the
device OFF region, and point B represents the lower limit of saturation or full ON. Switching could occur between these extremes, and
should the device current be driven beyond point B, the transistor
is operating as a 'saturating' switch. A non-saturating switch is
one in which the input signal causes operation to take place between
the cut-off region at point A and some other conducting point such as
C. Point C represents an ON condition but not a saturated ON.

Figure 9.17 shows the basic bipolar switch. When the emitter
junction is reverse biased, the base current is negligible, the
transistor is switched off, and only the collector leakage current
I_{CEO} flows. When the emitter junction is forward biased sufficiently,
I_B becomes large enough to drive the transistor into its saturation

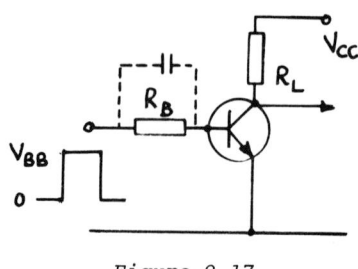

Figure 9.17

region. The collector current is then at its maximum (saturated or bottomed) value and V_{ce} is very small. The value of V_{ce} is usually designated $V_{ce(sat)}$. Then

$$I_{c(sat)} = \frac{V_{cc} - V_{ce(sat)}}{R_L} \qquad (9.5)$$

and the least value of I_B to produce saturation is

$$I_B = \frac{I_{c(sat)}}{h_{fe(sat)}} \qquad (9.6)$$

The transistor switch is, of course, imperfect for $V_{ce(sat)}$ is finite in the ON-state and I_{CEO} flows in the OFF-state. In silicon transistors, leakage is negligible but $V_{cs(sat)}$ is often not so; transistors selected for switching purposes should have a very low value of $V_{ce(sat)}$. In the FET, the ON resistance is given approximately by V_p/I_{DSS} and this compares favourably with the bipolar characteristics; but the OFF resistance of the FET is much superior, there being no leakage current. A figure of the order of 0.1 pA drain current would be typical for a MOSFET device.

The design of a transistor switch is then fundamentally centred on the saturation collector (or drain) current. For a bipolar device, the collector load R_L can be calculated from equation (9.5), taking $V_{ce(sat)}$ to be very small compared to V_{cc}. Then equation (9.6) can be used to find the required base current to produce hard saturation. This last value is usually multiplied by an 'overdrive' factor to make sure that the saturation condition is fully implemented; a factor of the order of 3 is normally used. Then from Fig. 9.17, if the input switching signal changes from 0 to $+V_{bb}$

$$R_b = \frac{V_{bb} - V_{be}}{I_B}$$

Example 9.4   A silicon transistor with $h_{fe(sat)} = 100$, $V_{ce(sat)} = 0.1$ V, is used in the switching circuit of Fig. 9.18. For an overdrive factor of 3, determine the required base current drive and the value of R_B if the input signal level varies from 0 V to +5 V.

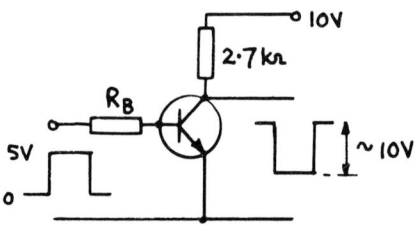

Figure 9.18

With the load resistor of 2.7 kΩ, $I_{C(sat)}$ $= \dfrac{10 - 0.1}{2700}$ A = 3.67 mA.
For $h_{fe(sat)}$ = 100, the base current for this collector current

$$= \frac{3.67}{100} \text{ mA } = 36.7 \text{ μA}$$

and for an overdrive factor of 3, this must be increased to 110 μA.
Hence, taking V_{be} = 0.6 V

$$R_B = \frac{5 - 0.6}{110} \times 10^6 \text{ Ω}$$

$$= 40 \text{ kΩ}$$

One point to notice is that the signal is inverted as always in
the common-emitter configuration; hence the output is low when the
input is high, and conversely.

The switching speed depends upon several factors, mainly the
shunt capacitance at input and output ports of the transistor, and
the effect of charge storage in the base region. Suppose a perfectly
rectangular pulse as illustrated in Fig. 9.19(a) is applied to the
base of a switching transistor. The transistor was in the OFF condition and the leading edge of the pulse turns it ON. Base current
flows immediately and follows the pulse rise as in diagram (b), but
there is not an immediate response in the collector current.
Electrons flowing from the emitter take a finite time in crossing
the base; some electrons move faster than others and take direct
paths, others take dispersed paths and travel more slowly. As a
result, collector current rises relatively slowly as diagram (c)
depicts. The rise time is taken as the time for the output pulse

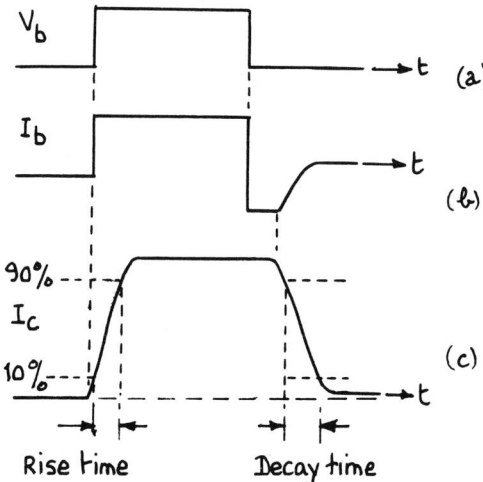

Figure 9.19

to rise from the 10% to the 90% levels of the peak pulse amplitude.

At the end of the input pulse the base current falls immediately but overshoots in the negative direction because of the high charge density built up at the emitter junction compared with the reverse bias condition. Thus for a period after the removal of the forward bias the junction still presents a low resistance. At the collector, where the heavy saturation current has been flowing for the pulse duration and the base-emitter voltage V_{be} has been of the order of 0.6 V, the collector is more negative than the base and emits electrons into the base region in addition to those from the emitter itself. This causes an excess of minority carriers (electrons) in the base region (where holes are the majority carriers) and the saturated transistor will not turn OFF until the base has been cleared of these minority carriers and the large charge built up there removed. There is thus a delay in switching to the OFF condition after the driving pulse has returned to zero. This is illustrated in Fig. 9.19(c). We are now faced with a dilemma: the base current overdrive which improves the switch-on rise time leads to the excess charge stored in the base region which results in a poor decay time at switch-off. If the base current is just reduced to

317

zero at the end of the input pulse, the base charge decreases simply because of charge recombination in the base and the base current return time is relatively long as Fig. 9.19(b) showed. If a reverse-drive factor can be introduced by the application of a reverse bias voltage at the end of the input pulse (in the same way as the over-drive factor is introduced at the start of the input pulse) then the stored charge will be removed from the base more quickly. This reverse base current will tend to cause the collector current to reverse, so the collector current will fall towards $-h_{fe}i_B$ rather than zero. The effect is illustrated in Fig. 9.20. The collector

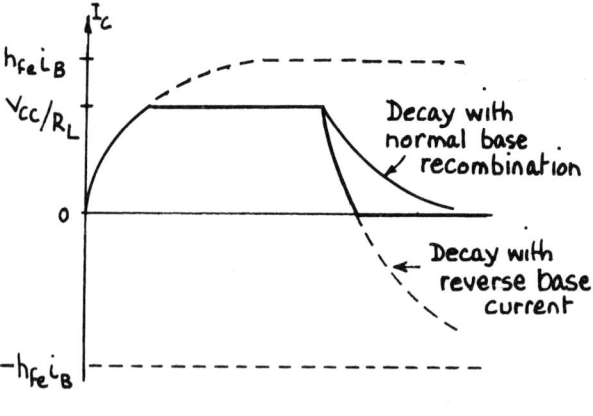

Figure 9.20

current cannot, of course, actually fall below zero because the stored base charge becomes depleted as I_c reaches zero; and the reverse base current must also stop because its source, the excess charge, has been dissipated. Thus, the introduction of a reverse drive factor speeds up the fall time in the same way that the over-drive factor speeds the rise time. By connecting a suitable capacitor in shunt with the base resistor R_B of a switching circuit (see Fig. 9.17) overdrive is secured at turn-on and reverse drive at turn-off. The charging current of this capacitor provides the extra drive at turn-on and the discharging current provides the reverse drive at turn-off. Such a capacitor is known as a 'speed-up' capacitor.

9.6.2   The astable multivibrator

The astable multivibrator (like the two other forms we shall discuss)
is an oscillator in the sense that its characteristics are the same
as those of the sinusoidal oscillators already encountered, that is,
a time-varying periodic waveform is generated without the introduction of any form of external input other than the d.c. supply,
and positive feedback is in operation. It is only in the shape of
the generated waveform that the astable oscillator differs from the
sinusoidal oscillator, or what we might call the 'continuous' oscillator. The circuit has 100% positive feedback ($\beta = 1$) and possesses
two conditions of stable equilibrium or stable states, hence the
name astable. In its basic form, shown in Fig. 9.21, each of the
transistors switch alternately from a conducting to a non-conduct-
ing state and vice versa, the period of such switching being

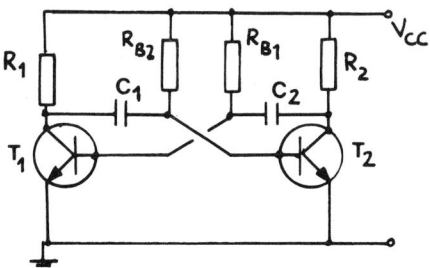

Figure 9.21

determined by the associated circuit elements. Thus transistor T_1,
for example, is ON part of the time and OFF during the remainder of
its period. In most cases, the ON condition is a saturated condition.

When the supply voltage is first switched on, currents flow in
the various circuit branches and charges are established on the
capacitors. Because the circuit arrangement is symmetrical, the
initial currents may well be similar, but dissymmetry in the component values and in the transistor parameters will quickly lead to a
cumulative difference in the two collector currents. This unbalance,
however slight, will cause one of the transistors to go into saturation and the other into cut-off. Assume that transistor T_1 has just
switched on. The left hand plate of capacitor C_1 will have been at
V_{cc} volts while the right-hand plate would have been at the base

319

voltage of T_2, that is, approximately zero, since T_2 is just switch-ing off. Hence C_1 will be charged to V_{cc} volts. When T_1 switches on, the collector falls to within about 0.1 V of the negative rail and C_1 transmits this change to the base of T_2 which then has a drop in po-tential carrying the base to that amount BELOW the negative rail. In the meantime an opposite change is taking place around T_2. Its col-lector has been at almost zero volts and now rises to V_{cc} as T_2 switches off. This change is transmitted by C_2 to the base of T_1, causing it to rise by the same amount. The circuit now 'relaxes' with T_1 fully saturated and T_2 cut off and C_1 beginning to discharge through base resistor R_{B2} and the collector-emitter path of T_1. This discharge path is shown in Fig. 9.22(a) with a simplified model at (b).

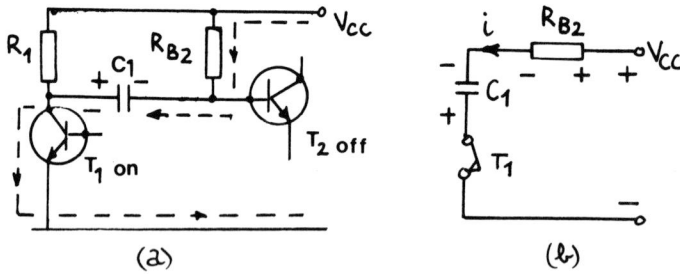

(a)   (b)

Figure 9.22

The voltage across R_{B2} will follow an exponential decay curve of the form

$$V_{RB2} = A \cdot \exp - \frac{t}{C_1 R_{B2}}$$

where A is a constant. When $t = 0$, $V_{RB2} = 2V_{cc}$ (very closely), hence $A = 2V_{cc}$.

Therefore

$$V_{RB2} = 2V_{cc} \cdot \exp - \frac{t}{C_1 R_{B2}}$$

T_2 will remain cut off until $V_{RB2} = V_{cc}$; let this occur after a time t_1 seconds. Then

$$V_{cc} = 2V_{cc} \cdot \exp - \frac{t_1}{C_1 R_{B2}}$$

Hence

$$t_1 = \ln 2 . C_1 R_{B2} = 0.695 C_1 R_{B2} \text{ secs}$$

In a similar manner, assuming a symmetrical circuit, the time T_1 is in turn cut off is

$$t_2 = 0.695 C_2 R_{B1} \text{ secs}$$

Hence the periodic time of the generated waveform $T = t_1 + t_2$ and the frequency

$$f = \frac{1}{0.695(C_1 R_{B2} + C_2 R_{B1})}$$

You should notice from this discussion that although only posi-tive direct voltages are supplied to the circuit, the switching action leads to voltage levels that go below the earth or negative line.

The waveforms at the bases and collectors of the two transistors are shown in Fig. 9.23. Here the broken lines give an indication of the ultimate capacitors voltages if the circuit would allow them fully to charge. From (a) in the diagram it can be seen that C_1 would tend to go through a total excursion of $2V_{cc}$; and from (c) that C_2 would do likewise. Actually, neither capacitor is permitted to do this, for when the voltage on either base gets slightly above zero, switching occurs.

The outputs at the collectors are good approximations to rect-angular waves with a mark-space ratio which is dependent upon the time constants of the cross couplings. Some care must be exercised in choosing the time constants, however, to ensure that the base resistors are still sufficiently low for the transistors to saturate at switch on. The departure from an ideal rectangular output, par-ticularly on the rising edge, comes about because as a transistor cuts off and its collector rises towards V_{cc}, the coupling capacitor

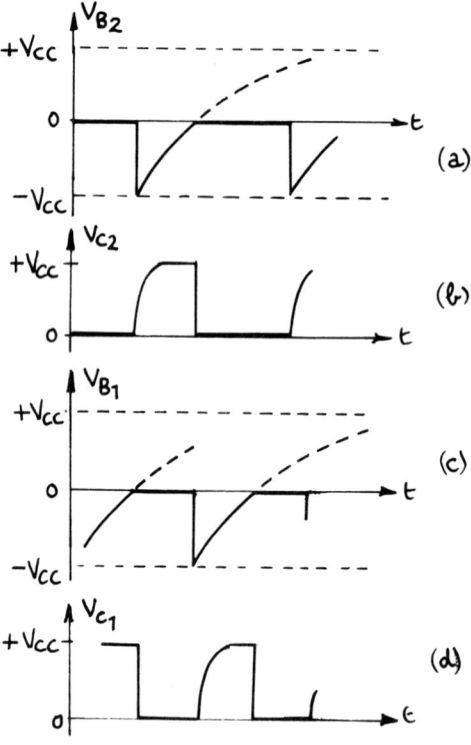

Figure 9.23

concerned acts as a very large 'stray' and has to be charged through
the collector load; look back at Fig. 9.22(a). This waveform im-
perfection can be rectified appropriately enough by the trick of
isolating the collector load from the capacitor by a series diode D
as shown in Fig. 9.24. When T_1, say, switches off, its collector
rises almost immediately to V_{cc} because the diode becomes reversed
biased and isolates C_1. Resistor R (which is equal in value to R_1)
now takes over the job of charging C_1 back to the V_{cc} rail. When
T_1 switches back on, the diode conducts and the sequence of operations carries on as already described. A second diode is of course
used at the collector of T_2 in the same way.

An operational amplifier may be used instead of the discrete
transistors to make a square wave oscillator; it has a number of
advantages which include easier construction, a minimal effect on

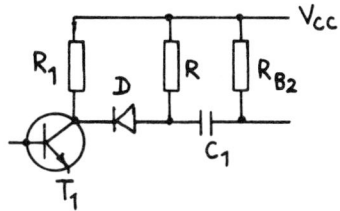

Figure 9.24

frequency by the output loading and, because of the high input resistance, smaller value capacitors may be used for a given frequency.
A circuit using the 531 amplifier is given in Fig. 9.25. The frequency is given closely by the expression

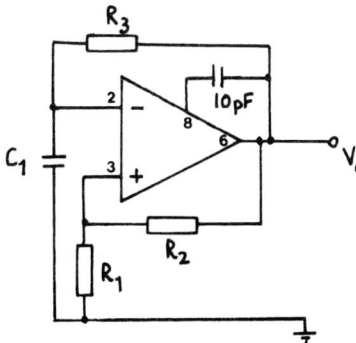

Figure 9.25

$$f = \left[2C_1 R_3 \ln\left(\frac{2R_1}{R_2} + 1 \right) \right]^{-1}$$

9.6.3 The monostable multivibrator

The astable multivibrator just discussed had coupling capacitors C_1
and C_2 in the feedback paths between the transistors, that is, the
coupling was a.c. coupling. The circuit therefore oscillated continuously between two 'quasi-stable' states whose durations were
determined by the time constants of the CR coupling components. In
the monostable multivibrator there is one a.c. coupling and one d.c.
coupling between the transistors; hence there is one stable state in

which the circuit will normally rest and a quasi-stable state into which it can be temporarily set by the application of an external trigger pulse. It is, therefore, fundamentally a pulse 'stretcher', the amount of stretching being under the control of the circuit designer.

In the basic monostable circuit of Fig. 9.26, transistor T_1 is biased so that it is in saturation, and transistor T_2 is biased normally at cut-off. Suppose now that a negative-going pulse is applied to the base of T_1 of sufficient amplitude to turn T_1 off. A sequence of events now follows which is best studied by reference to the simplified versions of the circuit shown in Fig. 9.27. Clearly, capacitor C determines the timing of the events. When T_1 is on and T_2 off, it can be seen from the first of the diagrams that C will be charged through R_2 to V_{cc}, the polarity being as indicated. When T_1 is switched off by the incoming pulse, T_2 turns on because of the collector of T_1 moving towards V_{cc} (i.e. going more positive), so that the base of T_2 goes more positive, and the circuit containing C

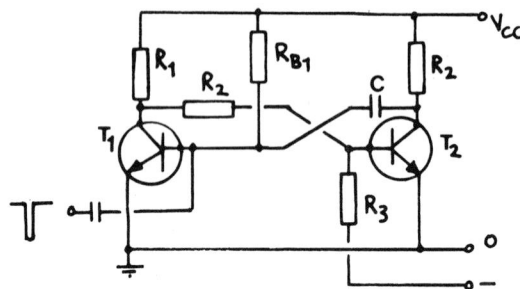

Figure 9.26

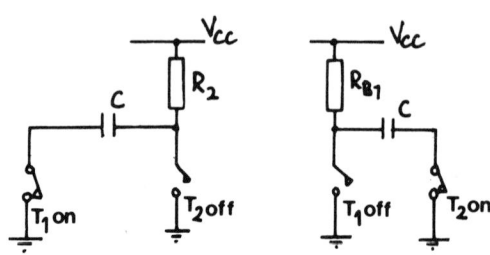

Figure 9.27

switches to the form shown in the second diagram of Fig. 9.27. Consequently, T_1 will remain off until the charge on C leaks way and the base of T_1 goes just above zero; at this point the circuit will rapidly revert to its stable condition with T_1 on and T_2 off.

An expression describing the performance of the monostable can be deduced by noting that the potential across C starts at V_{cc} and would eventually charge to V_{cc} of the opposite polarity if, as in the case of the astable circuit, it were not for switching taking place when the base of T_1 goes positive. Hence at a time t_1 ($= CR_{B1}$) the base voltage of T_1 is

$$2V_{cc}\{1 - \exp(-\frac{t}{t_1})\} - V_{cc}$$

You should particularly notice that during the quasistable state the circuit will not respond to any further negative input pulses which may appear at the base of T_1. It will, however, respond immediately to a negative pulse applied to the base of T_2.

The output is normally taken from the collector of T_1 in preference to T_2. The reason for this is because the pulse shape at T_1 is much better than that at T_2; the output from T_1 depends only upon the transistor response time, but at the other collector the coupling capacitor C has to charge from near zero potential to V_{cc} through the load resistor R_2.

An example of an integrated circuit form of the monostable is the 74121N and the basic circuit connections are given in Fig. 9.28. This circuit is much more sophisticated than the simple circuit of Fig. 9.26 and includes a Schmitt trigger system (to be described presently) which ensures that the operation is very reliable even if

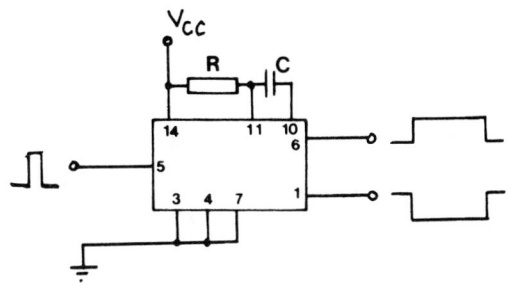

Figure 9.28

the input pulses are ragged or slow rising. Two outputs are available, each of equal duration given by the time $t = 0.695CR$, one being the inverse or negation of the other. In addition to pulse stretching which this application provides, the 74121 N can be easily adapted to provide pulse delay or the facility of manual triggering without the problems arising from contact bounce.

9.6.4   The bistable multivibrator

This circuit has two stable states and is capable of remaining in either state indefinitely until triggered by an externally applied signal pulse. For this reason, the circuit is commonly known as a 'flip-flop' of which there are a variety going under such names as the set-reset or R-S flip-flop, the data or D flip-flop and the J-K master-slave flip-flop. All of these have important applications in logic and counter circuits.

The simplest form of the bistable multivibrator is the discrete transistor form shown in Fig. 9.29. The capacitors shown in broken lines are speed-up capacitors; the coupling between the transistors is d.c. coupling by way of resistors R_3 and R_4. Input signals are fed to the transistor bases and outputs are taken from the collectors. Suppose T_2 is saturated and T_1 is cut-off. The collector of T_2 will be approximately at zero potential and the base of T_1 will be fixed by the divider R_4-R_{B1}; the choice of these values and the magnitude of the negative supply are such that T_1 is held off. Hence the collector of T_1 is relatively high and the values of the divider R_3R_{B2} are selected to make the base of T_2 positive enough to ensure saturation. The circuit then remains in this state until an external signal is introduced.

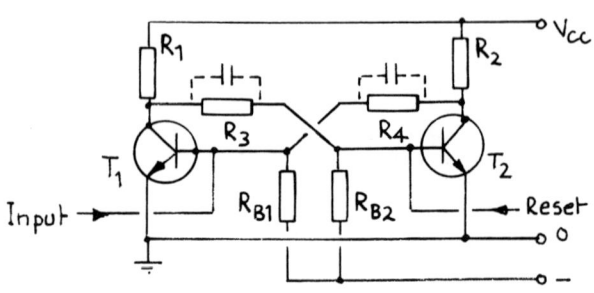

Figure 9.29

Suppose now that a positive pulse at the base input of T_1. This will override the negative bias and T_1 will conduct. The collector of T_1 will fall to about zero, carrying the base of T_2 negative. The collector of T_2 will correspondingly rise, augmenting and sustaining the positive change at the base of T_1. The circuit will therefore switch over and remain in the second of its stable states, even though the input pulse has ceased. The circuit can only be restored to its original condition by the application of a positive (reset) pulse to the base of T_2. This sort of operation indicates that the circuit can form the basis of a memory system; in a number of bistable circuits, each remembers the last signal received until another signal comes along.

An interesting aspect of the bistable multivibrator is that it can be fabricated from complementary transistors. A circuit is shown in Fig. 9.30. You should be able to analyse this for yourself by looking for the necessary phase shift around the loop, making a note of the fact that in this circuit the transistors are both SIMULTANEOUSLY on or off. Would speed-up capacitors help in this circuit, and if so where would you put them?

Example 9.5   A bistable circuit has $R_1 = R_2 = 1.5$ kΩ, $R_3 = R_4 = 1.6$ kΩ and $R_{B1} = R_{B2} = 1.4$ kΩ, see Fig. 9.29. The transistors have $h_{fe(sat)} = 50$ and the base currents when the transistors are saturated are 100 μA. Ignoring any leakage, calculate (a) the value of V_{CE} for either ON transistor, (b) V_{BE} for either OFF transistor, (c) V_{BE} for either ON transistor. Take $V_{cc} = 10$ V and the negative rail to be -4 V.

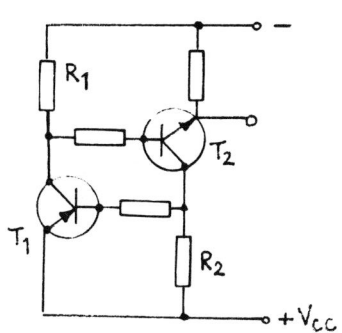

Figure 9.30

Assume that T_1 is ON and T_2 OFF; then referring to the simplified conditions of Fig. 9.31(a) we have, for a base current of 100 µA and an h_{fe} of 50

$$I_c = 50 \times 100 \text{ µA} = 5 \text{ mA}$$

Applying Kirchhoff across the diagram and working in kΩ and mA:

$$1.4I_1 + 1.6I_1 + 1.5(I_1 + 5) = 14$$

and from this we obtain

$$I_1 = 1.44 \text{ mA}$$

(a) The voltage drop across $R_1 = 1.5(1.44 + 5) = 9.65$

$$V_{CE(ON)} = 10 - 9.65 = 0.35 \text{ V}$$

(b) The voltage drop across the 1.4 kΩ

$$= 1.4 \times 1.44 = 2.02 \text{ V}$$

$$V_{BE(OFF)} = -4 + 2.02 = -1.98 \text{ V}$$

(c) V_{BE} for the ON transistor can be obtained by referring to the simplified circuit of Fig. 9.31(b). Again applying Kirchhoff across the circuit we have

$$1.4I_2 + (1.6 + 1.5)(I_2 + 0.01) = 14$$

and from this we obtain

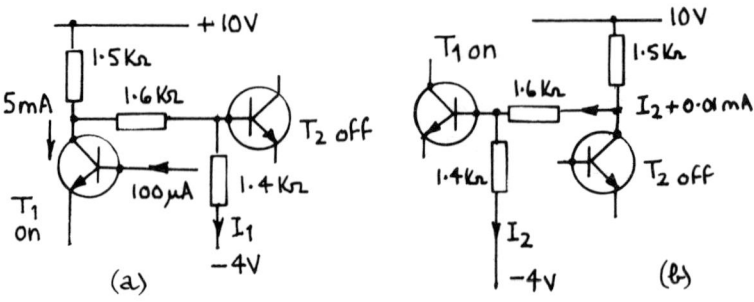

Figure 9.31

I_2 = 3.04 mA

The voltage drop across the 1.4 kΩ is then 3.04 × 1.4 = 4.25 V
and hence $V_{BE(ON)}$ = 4.25 - 4 = 0.25 V.

Example 9.6   In a laboratory experiment, a student assembled the
bistable circuit shown in Fig. 9.32, using the component values indicated. The transistors he used were known to have h_{FE} values of
about 50. When the circuit was switched on, it was found that it
oscillated freely in the manner of an astable design. Explain why
this happened and suggest a circuit modification which will enable
it to work in the way intended.

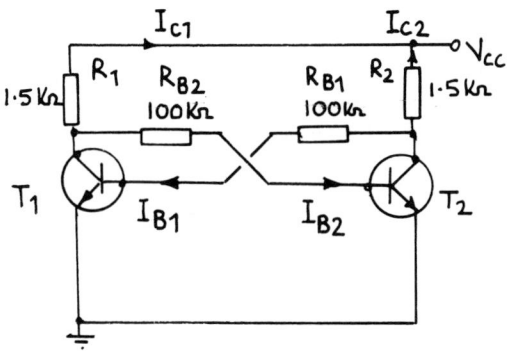

Figure 9.32

This circuit may look very basic, but it will nevertheless operate as a bistable when properly designed. The fact that the circuit
as made exhibited no stable state suggests that the transistors were
not being driven sufficiently hard into saturation. The base resistors are suspect in this case since clearly they determine the base
currents possible.

When a transistor is in saturation both junctions are forward
biased and hence the base and collector voltages are approximately
the same. Therefore if we assume that T_2, say, is saturated

$$I_{B2}(R_1 + R_{B2}) = I_{C2}R_2$$

and

$$R_1 + R_{B2} = R_2 \frac{I_{C2}}{I_{B2}}$$

But if T_2 is saturated

$$\frac{I_{C2}}{I_{B2}} \leqslant h_{FE}$$

so that

$$R_1 + R_{B2} \leqslant R_2 h_{FE}$$

or

$$R_{B2} \leqslant R_2 h_{FE} - R_1$$

Since $R_1 = R_2 = R$, say, and $R_{B1} = R_{B2} = R_b$, say, then the required condition for saturation to be possible is that

$$R_B < R(h_{FE} - 1)$$

From the student's design, $R = 1.5$ kΩ and $h_{FE} = 50$.

Hence for this circuit $R_B \leqslant 1.5(50 - 1) = 73.5$ kΩ. By using 100 kΩ resistors, the student prevented saturation and the circuit did not work as intended. 47 kΩ resistors would be adequate in this case.

9.6.5 The Schmitt trigger

This is a particular form of bistable multivibrator which is emitter coupled and which can be switched between two stable states by the 'amplitude' of the input voltage applied. Figure 9.33 shows the basic form of the Schmitt trigger circuit.

Assume T_2 is saturated because of the positive potential on its base derived from R_1, R_3 and R_4. The collector voltage of T_2 will be low and the voltage drop across the common emitter resistor R_E due to the collector current of T_2 will maintain the emitter of T_1 positive with respect to its base, hence holding T_1 off. The circuit will remain in this first stable state until an input voltage on the

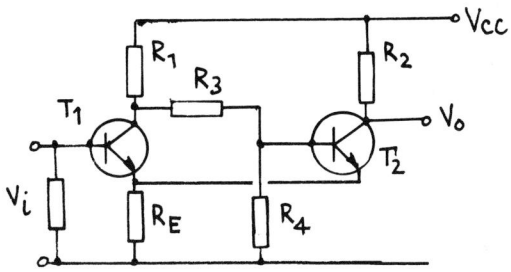

Figure 9.33

base of T_1 exceeds the combined voltage V_E (across R_E) and $V_{BE(ON)}$;
T_1 will then switch on. This voltage is known as the upper trigger
level, UTL. As collector current now flows in T_1, the fall in collector voltage reduces the positive potential on the base of T_2 and
at the same time the voltage across R_E rises. The combined effect
of T_2 base voltage falling and emitter voltage rising quickly leads
to T_2 being cut-off and the output at T_2 collector goes high.

The circuit remains in this second stable state until the input
voltage on the base of T_1 falls to a value which is lower than the
combined voltage V_E and $V_{BE(ON)}$. T_1 then switches off and the circuit rapidly reverts to the first stable state. Notice that the
voltage across R_E in this state is LESS than it is in the second
state as the magnitude of emitter current at which the lower trigger
level, LTL, is reached is lower than the corresponding value for the
UTL, the input voltage at the LTL usually not being large enough to
saturate T_1. In general, then, $V_{ON} > V_{OFF}$, and the difference
between these voltage levels is known as the 'backlash' or hysteresis
of the circuit. The action of the circuit is illustrated in Fig.
9.34. In practice, the values of the UTL and LTL can be varied or
made equal by a suitable choice of component values.

The Schmitt trigger has several useful applications, one of the
most important being its use as an amplitude comparator to mark the
instant at which some irregular waveform reaches a particular reference level. It can also be used as a pulse shaper. So long as the
input excursions are large enough to carry the circuit beyond the
limits of the hysteresis range, UTL - LTL, the output is a rectangular

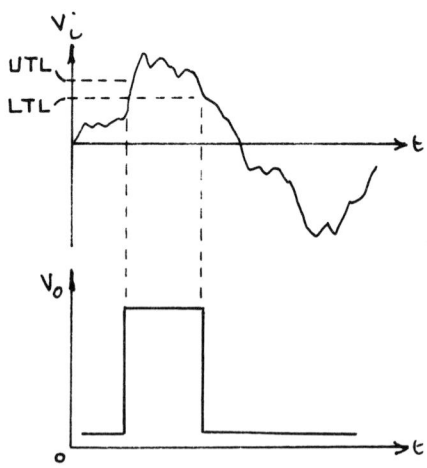

Figure 9.34

wave of the same frequency as the input but having an amplitude which is constant and independent of the input waveform. It is therefore useful as a 'cleaner' of ragged and noise-laden signals (and as a switch debouncer) before their application, for example, to a digital counter system.

Problems involving the Schmitt circuit can usually be worked by the direct application of Ohm's law and Kirchhoff. Consider an instant when T_1 is off, T_2 is on and the circuit is about to be switched; see Fig. 9.35(a). Neglecting I_{B2} in comparison with I_{C2} we have

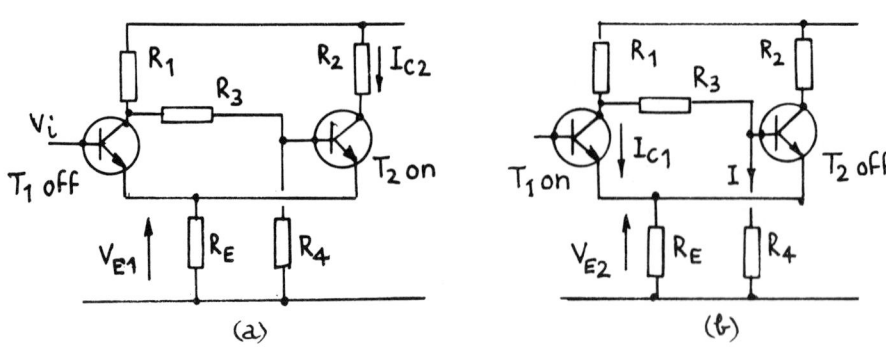

Figure 9.35

$$I_{C2}(R_E + R_2) = V_{cc} - V_{CE(sat)}$$

Then

$$V_{E1} = I_{C2}R_E = \frac{R_E}{R_E + R_2} (V_{cc} - V_{CE(SAT)})$$

$$\simeq \frac{R_E V_{cc}}{R_E + R_2}$$

that is

$$V_{ON} \simeq \frac{R_E V_{cc}}{R_E + R_2}$$

if we neglect $V_{CE(SAT)}$ compared with V_{cc} and ignore V_{BE}.

The divider chain R_1, R_3 and R_4 must be designed such that when T_1 is off, T_2 is saturated on.

After switching has taken place, the circuit condition becomes as shown in Fig. 9.35(b) with T_1 on and T_2 off. Reversion to the first stable state will occur when V_1 falls below the new emitter voltage V_{E2}, again neglecting V_{BE}. For I_{B1} negligible in comparison with I_{C1}

$$V_{E2} = I_{C1}R_E$$

and

$$I_{C1}(R_1 + R_E) + IR_1 + V_{CE(SAT)} = V_{cc}$$

Also

$$I(R_1 + R_3 + R_4) + I_{C1}R_1 = V_{cc}$$

Hence, neglecting $V_{CE(SAT)}$,

$$V_{E2} = I(R_3 + R_4)$$

Eliminating I between the above equations gives us

$$V_{E2} = V_{OFF} = \cfrac{\cfrac{V_{cc}R_E}{}}{R_E + \cfrac{R_1 R_E}{R_3 + R_4} + R_1}$$

Hence, for $V_{ON} = V_{OFF}$ (no hysteresis)

$$R_1 + \frac{R_1 R_E}{R_3 + R_4} = R_2 \qquad\qquad (9.7)$$

An integrated circuit Schmitt trigger is available in the 7413 chip which actually contains two identical circuits. The simplicity of using this integrated circuit is illustrated in Fig. 9.36 where it is used as an interface with digital systems to eliminate the effects of noisy signals. The 7413 has an hysteresis of about 0.7 V and so provides a clean rectangular waveform at the output even when the input is slowly changing.

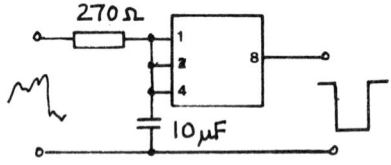

Figure 9.36

Example 9.7   In the circuit of Fig. 9.37, using the component values given, assess suitable values for the other components given that $h_{fe(SAT)} = 100$, $I_{C2} = 3$ mA and the UTL = 4 V.

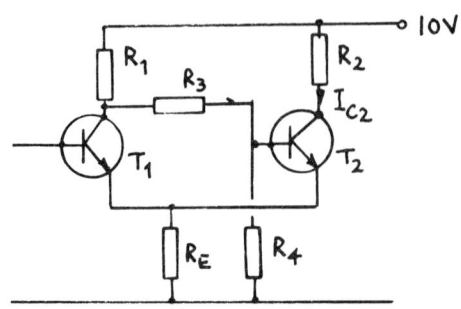

Figure 9.37

Since only an assessment of circuit values is required, the following assumptions may be made: $V_{CE(SAT)} = 0$, $V_{BE} = 0$ and $I_C = I_E$ in each transistor.

Then since $V_{ON} = 4$ V and I_{C2} is 3 mA, $V_{ON} = I_{C2}R_E$ and therefore

$$R_E = 4/(3 \times 10^{-3})\,\Omega = 1.33 \text{ k}\Omega .$$

Also

$$V_{cc} - V_{ON} = I_{C2}R_2;$$

hence

$$R_2 = 6/(3 \times 10^{-3}) = 2 \text{ k}\Omega.$$

Now

$$V_{OFF} = I_{C1}R_E;$$

hence

$$I_{C1} = 3/(1.33 \times 10^{-3})\,\text{A}$$

$$= 2.26 \text{ mA}$$

But

$$V_{cc} - V_{OFF} = I_{C1}R_2;$$

hence

$$R_1 = 7/2.26 = 3.1 \text{ k}\Omega$$

For $h_{fe(SAT)} = 100$, the minimum base drive needed for saturation of T_2 is

$$I_B = \frac{3 \times 10^{-3}}{100}\,\text{A} = 30 \text{ } \mu\text{A}$$

Hence the base current bleed through the resistance chain can be taken as, say, $5I_B = 150 \text{ } \mu\text{A}$. Then since

$$V_{ON} = R_4 \times 150 \text{ } \mu\text{A} \quad \text{and} \quad V_{cc} - V_{ON} = (R_1 + R_3)(I_B + 150 \text{ } \mu\text{A})$$

$$R_4 = 4/(150 \times 10^{-6})\,\Omega = 26.6 \text{ k}\Omega$$

and

$$(3100 + R_3)(180 \times 10^{-6}) = 6$$

so that

$$R_3 = \frac{6 \times 10^6}{180} - 3100 = 30.3 \text{ k}\Omega$$

Preferred values for the components could be R_E = 1.2 kΩ, R_1 = 3.3 kΩ, R_2 = 2.2 kΩ, R_3 = 33 kΩ, and R_4 = 27 kΩ.

The circuit could then be finalised from this initial set up.

9.7   A LABORATORY EXPERIMENT

This experiment investigates a phase-shift oscillator. Make up the circuit of Fig. 9.4, using two BC107 transistors and setting C_1 = C_2 = C_3 = 0.01 μF, R_1 = R_2 = 10 kΩ, R_3 = 12 kΩ, R_B = 47 kΩ, R_L = 680 Ω, R_E = 100 Ω decoupled by 100 μF. The V_{CC} supply should be 9 V, but adjustable.

Switch the circuit on and adjust R_E so that the whole of the resistance is decoupled i.e. the slider should be at the emitter end of the track. Connect an oscilloscope across the output terminals and examine the waveform obtained. Adjust the setting of R_E (or reduce V_{CC}) until a good sine wave output is obtained. Measure the frequency of the output wave using Lissajous figures or a digital frequency meter.

Using the formula for the frequency of this kind of oscillator, setting C = 0.01 μF, R = 10 kΩ, compare the theoretical frequency with that obtained in practice. Account for any discrepancy.

Adjust R_E (or V_{CC}) until the oscillation is JUST maintained. Now measure the gain of the amplifier section by removing C_3 and feeding a signal of about the same frequency as the generated oscillation from a signal generator into the junction of R_3 and R_B. Compare your figure obtained for the gain with the theoretical figure of -29. Account for any discrepancy.

PROBLEMS 9

1. Distinguish between a harmonic and a relaxation oscillator.

2. 'All oscillators are phase shift oscillators.' Comment on this assertion.

3. A student answering an examination question wrote: 'An oscillator is an unstable amplifier'. Comment on the aptness, or otherwise,
of this definition.

4. A feedback system has $\beta = 0.005\underline{/0^{\circ}}$. Define the gain condition
necessary for oscillation in the system.

5. For frequencies less than $f = 318$ Hz, the gain of an amplifier is
described by $A = 0.2\pi f\underline{/180^{\circ}}$. What value of feedback factor is required for oscillation to occur at a frequency of about 240 Hz?

6. The phase shift between the input and output terminals of a certain ladder network made up of capacitance and resistance is expressed by

$$\phi = \frac{270^{\circ}}{1 + f/500}$$

where f is the frequency in Hz. This ladder is used in the phase
shift circuit of Fig. 9.4. What will be the approximate oscillation
frequency?

7. What are the limits of phase retard and phase advance in the CR
circuits of Fig. 9.2(a) and (b) respectively? Sketch appropriate
phasor diagrams for angles of phase shift approaching 90°. What do
you deduce from these diagrams?

8. Define phase stability. Show that a necessary condition for
phase stability in a tuned oscillator is that the circuit Q should
be large.

9. Verify that the frequency of the generated square wave in a symmetrical astable multivibrator (Fig. 9.21) is given by $f = 1/1.38CR$,
where $C = C_1 = C_2$ and $R = R_1 = R_2$.

10. Fig. 9.38 shows part of a multivibrator circuit. What is (a)
the time constant of this circuit, (b) the voltage across R_B when
T_2 just switches on, (c) the charge on C is that instant.

11. The components of the feedback network of the Wien oscillator
of Fig. 9.8 are $C_2 = C_3 = 10$ μF, $R_6 = R_7 = 4.7$ kΩ. What will be the

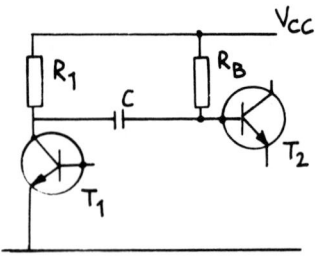

Figure 9.38

approximate frequency of oscillation? Why is this solution only
approximate?

12. The transistor of Fig. 9.39 is considered to represent an ideal
switch in that V_{BE}, $V_{CE(sat)}$ and I_{CBO} are all zero. If h_{FE} for this
transistor is not less than 70, calculate the required value of V_{CC},
R_1 and R_L if the input pulse train amplitude is 3 V, V_o = 12 V peak
and $I_{C(sat)}$ = 15 mA.

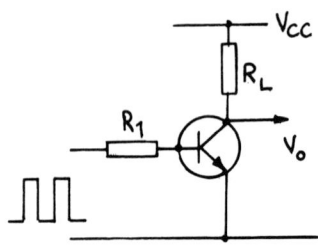

Figure 9.39

13. Derive expressions for the frequency of oscillation and the
smallest value of mutual inductance M necessary to maintain oscillations in the circuit of Fig. 9.13(a). You may assume that h_{re} and
h_{oe} are negligible and that the circuit Q is large. State any other
assumption you may make.

    If L = 1 mH having a resistance of 4Ω, and C = 500 pF, estimate
the frequency of oscillation and the necessary least value of M,
given that h_{ie} = 2 kΩ, h_{fe} = 50.

14. The transistor used in the circuit of Fig. 9.40 has parameters

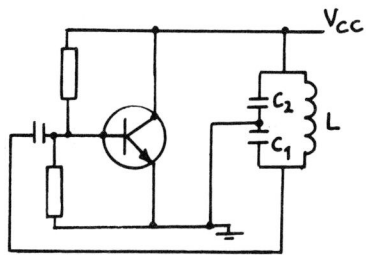

Figure 9.40

h_{ie} = 1.4 kΩ, h_{re} = 4 $\times$ 10^{-4}, h_{fe} = 50, h_{oe} = 150 μS. Determine an expression for the frequency of oscillation of the circuit, neglecting the effect of the bias components. Justify that this expression may be approximated to that of the tuned circuit alone, using the parameters given. If L = 200 μH and C_1 = 1500 pF, what is the theoretical maximum frequency of operation, assuming that the h-parameters remain constant?

15. Derive an expression for the current or voltage transfer gain of the network shown in Fig. 9.3. Thence indicate how this network could be used in conjunction with a bipolar or FE transistor to form an oscillator. Sketch a circuit diagram and establish the requirements of gain and phase shift in the amplifier for the circuit to function as an oscillator. Comment on the impedance matching requirements of the network when it is connected appropriately to a bipolar or FE transistor.

Given that R = 10 kΩ and C = 0.01 μF, obtain a figure for the frequency of oscillation.

16. In the circuit of Fig. 9.41 show that there is only one frequency at which the phase shift is zero, and derive an expression for this frequency in terms of the circuit components.

Sketch a circuit of an oscillator which makes use of the above network in the feedback path, and deduce a figure for the minimum theoretical gain of the amplifier for oscillations to occur, given that R_1 = R_2 and C_1 = C_2. Name this type of oscillator and make brief notes on any advantages the circuit has when utilised as a

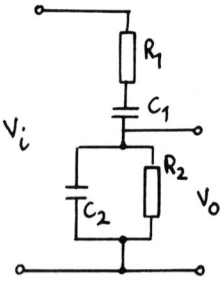

Figure 9.41

'variable' frequency oscillator.

A two-stage transistor amplifier has effective input and output resistances of 1.2 kΩ and 2.5 kΩ respectively. It is used as an oscillator of the type discussed above and the values R_1 = R_2 = 15 kΩ, C_1 = C_2 = 4000 pF are used in the feedback network. Obtain estimations for (a) the frequency of oscillation, (b) the amplifier gain when the oscillations are stable. (Note: the answer to this second part is NOT 3!)

17. In the Schmitt circuit of Fig. 9.33, describe the action of the circuit to a sinusoidal wave applied to the base of T_1. Explain what is meant by the hysteresis of the circuit and mention two practical applications to which the circuit might be put.

You are called upon to design a circuit of this form having the following specifications:

Supply voltage V_{CC} = 10 V
Collector currents to be of the order of 3 mA.
Upper trigger level (V_{on}) = 4 V
Lower trigger level (V_{off}) = 3 V

Transistors are available having $h_{fe(sat)}$ = 100.

Making note of any approximations or assumptions used in your trial design, assess suitable experimental values for the five resistors used in the circuit. Indicate the nearest preferred values you would initially use in your circuit.

Solutions to selected numerical examples

2. PROPERTIES OF SEMICONDUCTOR DEVICES

6. $I_{CEO} = I_{CBO}(1 + \alpha_E)$

8. -1.2 mA

10. 4.98 mA

13. P = emitter, R = base, Q = collector. The emitter is p-type.

16. 1.005 mA, 4.95 μA, 0.99, 0.505 mA.

4. LARGE SIGNAL AMPLIFIERS

6. 75

12. 34.5%, 7.2 W

13. 38.5 V, 0.39 A; 6.5%; 480, 2 stages.

14. (a) 11.25 W (b) 3.3% (c) 14.5 V (d) 24.6 W (e) $4.67^{\circ}C/W$

15. 25 W. Case $118.75^{\circ}C$; washer $88.75^{\circ}C$

16. $5^{\circ}C/W$

18. 6.95 W

19. (i) $1.2^{\circ}C/W$ (ii) 25 W (iii) 200 cm^2

20. (a) 5 W (b) 32 dB (c) 32%

21. (i) 2 mW/$^{\circ}C$ (ii) $0.5^{\circ}C/W$. S $\simeq$ 27

5. SMALL SIGNAL MODELS

6. Z_i = 1 kΩ, Z_r = 10 Ω, Z_f = -1 MΩ, Z_o = 10 kΩ

12. (a) 33 (b) 292

13. A_i = -16.7, A_v = 417, A_p 38.4 dB

14. 5 kΩ

16. (a) 2.5 mS, (b) 1.5 mS

17. (a) 3.1 mS, (b) 3.0 mS

18. 3.375 mA; 4 mV. 300Ω

19. g_m = 2.6 mS, r_d = 10 kΩ, μ = 26

20. 7.7 kΩ. A_v = -500, A_i = 78

6. SMALL SIGNAL AMPLIFIERS

11. ≈ 0.023 μF

12. 40

14. 50 kHz

16. ≈ 25 Hz; 34.5 kHz

7. FEEDBACK

3. ≈ 13; 1.1%

4. 6600; 0.0098

5. β = 0.08.   20 Hz - 1.24 MHz

7. 86/-122°; negative feedback

8. At 1 kHz, A' = 50/0°. May be unstable at 10 Hz when β|A| = 2.

10. (a) + 1.02 dB   (b) -0.13 dB

14. 20 dB

15. R_L = 220 kΩ, R_o = 200 Ω

18. β = 0.0019, A' = 500

20. 1/6; $\sqrt{3} \times 10^5$ Hz

8. DIRECT-COUPLED AND OPERATIONAL AMPLIFIERS

10. 18 μs.

12. 100 kΩ; 101.1 kΩ.

14. -31.5 V

16. 4.75 kΩ; 4.70 kΩ

18.
$$v_o = \frac{R_L h_{fe}}{h_{ie}} \cdot \left[\frac{\delta v_1 R_E (1 + h_{fe}) - \delta v_2 R_E (1 + h_{fe})}{h_{ie} h_{ie} + 2R_E (1 + h_{fe})} \right]$$

Now substitute $\delta v_1 = v_m + v_d$

$$\delta v_2 = v_m - v_d$$

19. 59.8 to 60 dB; ≈ 3.1 MΩ; ≈ 650 Hz.

9. OSCILLATORS AND WAVEFORM GENERATORS

4. 200/0°

5. 0.0067/180°

6. 250 Hz

10. (a) $R_B C$ (b) V_{CC} (c) Zero

11. 3.3 kHz

12. V_{CC} = 12 V; R_1 = 17.6 kΩ; R_L = 800 Ω

13. 228 kHz, M = 0.08 μH

14. 4.7 MHz

15. 650 Hz

16. (a) 9.025 kHz; (b) 15.8

17. Assumptions: $V_{CE(sat)}$ = 0, I_C = I_E, V_{EB} = 0. Then suitable preferred values are R_E = 1.2 kΩ, R_1 = 3.3 kΩ, R_2 = 2.2 kΩ, R_3 = 33 kΩ, R_4 = 27 kΩ.

Index

—